heytesbury stud

AN AFFAIR OF THE HEART

The Romance of Heytesbury Stud

heyte

heytesbury stud

AN AFFAIR OF THE HEART
The Romance of Heytesbury Stud

ETHNÉE HOLMES À COURT
with TANGEA TANSLEY

GARY ALLEN G

Published by
Gary Allen Pty Ltd
9 Cooper Street
Smithfield
NSW 2164
ABN: 34002 793 160

First edition published 2003

Printed by Griffin Press

Cover design and layout: Sharon McGrath

Cover photo: Horse whispering—With my much-loved Silver Smile. By Derrick Nicholas.

Packaged: Media21 Publishing Pty Ltd

National Library of Australia Cataloguing-in-Publication data:
Holmes à Court, Ethnée
Heytesbury Stud,
biography

ISBN 1877082-31-7

Disclaimer: The events and individuals have been described as I remember them.
If there are any inaccuracies or errors, please accept my apologies.

I dedicate this book to Ronnie, my beloved husband of twenty-one happy years. He died peacefully at Heytesbury on 27th August, 1999.

And God took a handful of southerly wind,
blew his breath over it...and created the horse.
Bedouin legend

Contents

Introduces Heytesbury as it is today on
the colourful occasion of Stallions Day

My first sight of the run-down farm
that Robert planned to make into 'the
most fabulous thoroughbred stud'

Work begins on tidying up the
property, buying the first crop of
broodmares

Witnessing the birth of Tixall's colt, the
first foal to be born at the stud

Heytesbury's foundation stallion arrives
to take up his duties

Acknowledgements

The writing of this book has been very important to me for two major reasons. Firstly, it documents the thrills and occasional spills involved in building a premier horse stud from a scratch start. My life and that of Heytesbury Stud have been so closely entwined that few others have any knowledge of the events I describe and it is important that the history of this stud not be lost. Secondly, this book is dedicated to my fourth husband Ronnie, sadly no longer here in person, but maybe in spirit. Certainly his words of encouragement reach me on a daily basis. He was the dearest of men and I miss him a great deal.

My thanks are due to my computer tutor Mike Finn who, until his recent relocation to New Zealand, not only extricated me from the sometimes overwhelming frustrations of the computer age, but who also took many of the photographs for this book. I am most grateful to him on both counts. In Mike's absence, thanks, too, to Derrick Nicholas who has been equally patient with my computer struggles and has also taken a considerable number of photos of the stud, some of which appear in the following pages.

I am very grateful to my friend of many years, Marjorie Charleson, for introducing me to Tangea Tansley, my ghost writer for this project. Tangea and I have similar backgrounds in that we both spent a significant proportion of our lives in what was then Rhodesia and in South Africa and importantly we both love animals. We worked closely together for almost a year in the writing and production of this book and I have fond memories of the many happy hours we spent in my cottage at Heytesbury, our work punctuated by stories and recollections, lots of laughter and many cups of tea.

The friendship of my carer, Jacque Meyer, who is always so unfailingly helpful and patient, needs also to be acknowledged. I often wonder how she manages to keep so cheerful, particularly during the foaling season when she is up most of every night. I appreciate and value her help enormously.

Lastly, a big thank you to all my wonderful friends. So many of you have asked when my next book was coming out and here it is. I hope you enjoy the story of Heytesbury Stud as much as you did *Undaunted*.

Ethnée Holmes à Court
Heytesbury Stud
October 2002

Foreword

Having been deeply honoured with the request to write this foreword, I am delighted to have the opportunity to pay tribute to a most remarkable woman in Ethnée Holmes à Court.

Those who have read her first book *Undaunted* will appreciate her resilience, determination and spirit illuminated by a strong sense of humour, all qualities which have brought her through eighty-seven years of a challenging life.

I was privileged to meet Ethnée through a mutual friend not long after her arrival in Western Australia nearly thirty-five years ago and was one of the early visitors to the 'farm' shortly after it had been purchased by Ethnée's elder son Robert. This was about the time referred to in the second chapter of this book when there was "barbed wire" aplenty and not a great deal to enthuse about beyond a small bush of red roses.

However, Robert's skills in farm and pasture management had been well-honed at the famed Massey Agricultural College in New Zealand and he applied these practices to great effect on the fledgling Heytesbury Stud.

He also looked to the New Zealand thoroughbred in establishing his breeding empire when he purchased the four-year-old grey stallion Silver Knight after he had won the 1971 Melbourne Cup. Silver Knight was by the champion Australian sire of the late Sixties, Alcimedes, from Cuban Fox, by another champion New Zealand-based sire Foxbridge and a daughter of the 1966 Broodmare of the Year Chubin by the great Nizami. Truly a most distinguished pedigree of classic and cups winners. At the same time he started to import a band of well-bred mares, the end results of which are still apparent today in producing winners off the stud. It was the pairing of Silver Knight with one of these foundation mares, Brenta, that produced Black Knight, winner of the 1984 Melbourne Cup for Heytesbury.

Although he went on to become one of Australia's wealthiest entrepreneurs, my personal feeling was that Robert's true passion was for his horses. Certainly on the occasions of his visits to Heytesbury in those early days when walking and talking in the paddocks among the broodmares and foals there was little evidence of the boardroom tycoon.

The fact that Heytesbury has not only recovered from the premature death of its founder, but is again set to be a major force is in no small way due to the sound foundations, in the most literal sense, laid by Robert. However, there is also no doubt that his love and respect for the horse was inherited from his mother who was an accomplished horseperson from about the time she could walk.

Ethnée's story of Heytesbury is a warm and vibrant account of dreams realized, of her own wonderful romance, and of some aching sadness—in all a true story of some remarkable people, not the least of whom is Ethnée herself.

Marjorie Charleson
Western Australia 2002

CHAPTER ONE

Stallions Day

It's Stallions Day at Heytesbury Stud. One by one the splendid animals are to be brought out of their stables and paraded in front of a crowd that has been gathering for the past two hours. The horses know something special is going on and they are hustling and stamping in their boxes, senses as keen as the needle on a radar screen. Perhaps it reminds them of their racing pasts: the anticipation, the colour, the people, the fluttering fringes of the marquee at the far end of the parade ground.

Lord Dane has been using both hind hooves on the inside of his stable door creating strings of sharp staccato sounds which shoot through the assembly like a flurry of Chinese fireworks. He is new to stud work, excited by his first serving which took place only the day before, further roused by the crowds. A young woman carrying a little girl stands outside his stable. The child clings nervously to her mother tucking her head into the woman's neck as the bay surges back and forth behind the stable door. I hear the mother reassure her, 'Don't worry, darling, he can't get out'. I admire the calmness of her tone, the confidence in her voice. After a lifetime spent around horses, I wouldn't be quite so sure.

But I am excited, too. This is the second day of spring with Stallions Day, Father's Day, the twelfth anniversary of my son Robert's death, all packed into one afternoon. Three hundred invitations sent out, three hundred acceptances received; the stallion books already well on their way to being filled. So much to think about. So many people I know to greet, talk with, introduce to one another.

While Stallions Day may resemble a race day in terms of a sense of eagerness so powerful that it almost throbs through the air, it is quite different outwardly. There are no binoculars or worked-over programmes; no saucy hats or particularly elegant clothes. In fact most of the women wear trousers, many of the men are in jeans. This is decision-making time behind the scenes, a gathering of experts, each with critical eye and figurative pencil at the ready. One concession to the celebratory nature of the occasion are the glasses foaming with champagne. But the atmosphere is no less electric for all that. This is what the horses pick up.

My youngest grandson Paul lifts the microphone and from a position at the top of the parade area opens the event. Someone at the back calls out that he cannot hear properly. Nonplussed Paul calls back, 'This is as good as it gets.' The public address system is adjusted and everyone settles to enjoy the occasion. I look at Paul and my heart glows. His mother Janet and sister Catherine are both here today. 'To show support for Paul and for the stud,' Catherine told me.

Carry A Smile is the first to be paraded around the ring. He is sixteen now and his once-dark coat glistens with shades of dappled silver in the sunlight. His mother is Silver Smile, the beautiful grey dam produced by Silver Knight, the first stallion to arrive at Heytesbury. His sire was none other than our great Haulpak. His ears shoot forward and his eyes glint in recognition as he passes me. He shakes his head in its shining brass and leather head-collar and a tremor shimmies across his withers and along his flanks. Normally I have carrots for him. Not today. Just a smile and memories between us.

The crowd is quiet. Intent. Anticipating. Nobody is talking about the weather any more. The forecast was for showers, and the black clouds crouching low over the ranges did indeed look rather menacing earlier. But now the only clouds in the sky are high and fluffy and moving away to the south. Three hundred people crowded into the marquee had been one of my fears. We need rain badly this year but, please, not this afternoon, not for the next two or three hours.

As I scan the sky, my gaze fastens on two Wedge-tailed Eagles playing tag on upswings of air along the top of the ranges. Yesterday they were spotted feeding on a dead kangaroo. Now they are soaring in huge swooping circles high above us, doubtless looking for another feed. Despite the lack of rain the hills unfolding above us are green right to the horizon. Beyond the marquee the paddocks, too, are lush with pasture. Here and there the stud's new mothers hang their heads low over tufty-coated foals. Small breezes tease the leaves of the grand old gums that line the avenues that stretch away from us in all directions. Heytesbury Stud is at its best.

By the time Lord Dane is brought out, he is less agitated. He shows himself surprisingly well, proving that all he wanted to do all along was to join in the celebrations. I am struck again by the power in his hindquarters and the sense of challenge that simply pours out of his big body. At 15.3 hands, he is not a particularly tall horse, but he has a presence that makes him seem larger than he really is. No photo can do justice to Lord Dane: a camera would be hard put to capture the surging vigour of this horse. He was sired by the renowned Danehill, the U.S. stallion which produced 150 Stakes winners including five Golden Slipper firsts. Like Haulpak, he was prevented by early injury from reaching his potential on the track—a burning shame I think as he passes in front of me. He was born to race. Perhaps he will produce top stud fees, set this season at $4,000. Again, memories take over as I think of the tentative $250 fee we first charged for Haulpak, which over the years soared to what remains a Western Australian record at $15,000. I watch Lord

Dane being led back to his box, quite passive now. How long will it take for him to topple the great Haulpak's record?

The last horse to be introduced is our shuttle stallion Second Empire which flew over from Ireland four days before. Before he was retired to stud, he was a champion three-miler in his homeland. These days he spends half the year serving at Coolmore Stud, the other half here. This is another stallion with an impeccable international pedigree. He is Western Australia's first and only shuttle stallion and his book tends to fill very quickly. It is almost as though he remembers Heytesbury because he has settled so easily this time. His handler, Christoph, travelled with him from Ireland again this year and is never far from his side. As he leads the stallion out, it is difficult to say who exudes the most pride: the handler or the glossy dappled bay horse. Like others of his calibre, this stallion tends to stamp his foals with his own colour and markings. Eleven months from now there will be another set of foals with the same distinctive star and, hopefully, the stamina of their sire.

My gaze shifts from the horses to Paul who is reaching for the microphone once again. Paul Holmes à Court: both so like his father and yet so unalike. This time it is *my* heart that overfills with pride. Only eighteen months after he took control of the stud, he was promoted to the position of Heytesbury's Chief Executive Officer. Today he is here to open and champion this event which he has tightly sandwiched between a business trip to Asia and a visit to the stations up north. He is only twenty-eight years of age, but he has looks, presence and charm. Like Robert he is quiet and self-assured; at the same time he is one of the most diplomatic people I have ever met. He is a very feeling person who has already built up a great degree of trust in his young life—it's obvious from his smile which seems to come from the very centre of his being.

Something a little different is to happen this year. The inclusion of dogs along with the horses. As a sponsor of the Guide Dogs

Association I am to present a cheque from part of the proceeds of the sale of my first book *Undaunted* to the president of the Association for the Blind. She is here with four of her members, with their golden Labradors harnessed and complacent beside them. Paul promised that he would match whatever I raised, but I don't think he quite expected my efforts to reach $10,000! This means that the total of the cheque we are handing over is $20,000, enough to train a young puppy to guide dog status. Of course I have named the pup already. We have decided, no matter what the sex, to call it Ronnie after my late husband.

Never one to be constrained by tradition, Robert frequently departed from the convention of naming horses after their sires and dams. So we once had a horse he called Templeman after a lawyer; a mare acquired during the bid for control of Ansett became Lady Pilot; and our noble black stallion was dubbed Haulpak after Bell's huge haulage trucks of the north. And so it went on until in time the practice carried over to my own four-legged loves: Lara the Rhodesian Ridgeback dog and Zhivago the Russian Blue cat. Soon Ronnie the Labrador will be added to the list. Just one instance of the many ways in which, twelve years after his death, my dear Robert's presence is as strong as it ever was. My sight rests on a point beyond the clouds. Very often these days I find myself raising my eyes upwards to ask a question or gain affirmation for something I am about to do.

The crowd is thinning. The more serious party people are drifting in the direction of the marquee for coffee and cake or clustering at small tables to sip at glasses of Vasse Felix wine. Others have strolled off down the avenues to look at the new foals. Nine of our mares have foaled already this season: my two mares—Wild Rumour and Mercurial Madam—on the same night. Despite a serious accident two weeks ago, Mercurial Madam has given birth safely to the most beautiful filly. Although the horses aren't officially named until they are yearlings, for the moment I am going to call her Mercurial's Dream.

No. As usual, I am running ahead of myself. This is the story of how Heytesbury Stud came to be and I must start at the very beginning.

It is exactly thirty years ago that I first saw the rundown farm that was to make its mark as a leading stud. This is how it happened...

CHAPTER TWO

Barbed Wire and Red Roses

However long it may have been in the making, change often seems to take place quite suddenly. One moment life seems to have been forever the same, the next it is quite different.

For instance, the rather cool autumn day of May 3, 1971 appeared to be just one more day in my working life as a bookkeeper for MRH Holmes à Court & Co—Barristers and Solicitors. For the past four years—ever since my arrival in Perth from Rhodesia—I had been working for my son Robert, a pastime that had enough in-built excitement to be worth a book in itself. Everyone who has ever worked for Robert has a tale to tell and I am no exception.

But for some time now, life had been progressing quite smoothly. This was during the early years of Robert's practice as a barrister and we were working out of Pamos House in Adelaide Terrace. In retrospect I can't help thinking that Robert did well to surround himself with such a rock-solid support base. At that time the office staff consisted of three: Janet, his wife, who was responsible for the titles and conveyancing work and who was devoted to him until his death; his secretary Val who worked loyally and efficiently for him until her retirement twenty-five years later; and myself, his mother. Except he never called me 'mother'; it was always Ethnée.

My role within the firm was that of bookkeeper. Originally I had set out to do a course in journalism. I enjoyed writing and could see myself as a journalist, but it had been Robert who stopped me. 'What do you want to be bothered with that for?' he had asked. 'You would be much better off doing a course in bookkeeping.' Robert's word was law and so I, admittedly rather reluctantly, dropped out of the journalistic course, learned the rudiments of bookkeeping and worked for my son instead.

At least it provided me with a secure job and a good living and we three support staff got along very well. Each day was filled with gossip and chatter as we went about our various tasks. So much so that it upset Robert who had a large office next to ours and who could evidently hear the laughter through the wall separating the two rooms. No sooner had he decided that we were having far too much fun to be seriously working than we were separated. With some justification, he decided that it was I who was the perpetrator and thus the unsettling influence amongst the three and so, without further discussion, I was moved to a back office all by myself.

This particular morning, Robert entered my office and walked up to my desk in his usual relaxed and unhurried manner. I looked up and waited. This was of course not unusual. With Robert you could wait a long time for a sentence. There was no point in attempting to hurry him because he simply didn't hear you. He glanced at the papers on my desk and took several short puffs at his cigar before stubbing it out in the glass ashtray in front of me. Then he spoke in the sudden way he had that often followed a prolonged pause.

'I promised you that we would have a farm and horses again one day. I knew you would never settle here properly until we did. Well…I've bought one.'

There is a short space of time that closely follows shock, when everything is presented in exaggerated relief and the silence itself is loud. In the moments after his statement I had no heartbeat, no breath. There was only Robert towering over me, looking even taller than usual, the picture on the wall behind him irritatingly askew. The only thing moving in the room was a thin curl of smoke from the dying cigar.

I have always been hopeless at hiding my feelings. There is too much enthusiasm in my character for that—and somehow not enough time in life. In this, Robert and I could not have been more opposite. He watched with just the trace of a smile on his face as the spark of excitement crackled right up through my body and burst out all over my face. I think I leapt out of my chair. I certainly clapped my hands.

A farm. Animals. Horses. I felt like crying with the sheer unexpectedness of it. Those of you who have read my first book *Undaunted* will know of my upbringing and life in Africa and my deep love of horses and riding. You will understand how I felt. My heart felt as though it were beating out its life's rhythm on happiness alone.

It seemed we were to visit the farm that afternoon. To all my questions, Robert would only say, 'Wait and see'.

And so I had no option but to wait. Impatiently. All that focused energy. It was impossible to concentrate. Rows of figures danced to the new thump of my heartbeart. I could feel my pulse pounding away in the pads of my thumbs. In place of the office desk, the dirtied ashtray and the yellow-cream walls, there was a grand old homestead, stables, paddocks…and horses, horses, horses.

I closed my eyes and tried to imagine it into being. I had very little to go on. Where would it be in relation to Perth? North, south or east? How close? Was it inland or near the ocean? Would it be flat and swampy or hilly and undulating? Would the soil be sand or gravel? And what sort of buildings would there be on the property? 'A farm', Robert had said. What sort of farm,

I wondered. The alarming thought occurred to me that he had bought the property for himself and Janet and their first two young children, Peter and Catherine. I tried to repeat his exact words in my mind: *I promised you that we would have a farm and horses again one day. I knew you would never settle here until we did. I've bought one.*

It sounded as though I were included but nevertheless there was the beginning of uncertainty.

But this didn't stop me from becoming more and more excited. How slowly the hands of the clock moved that day.

Eventually, however, two o'clock did arrive with Robert behind the steering wheel of his Rolls just as uncooperative as he had been earlier in answering my questions. Aside from telling me in a few dry sentences that Janet had found the property for sale in the course of her conveyancing work, that he had subsequently seen it and bought it, all he would do was repeat what he had said earlier: 'Wait and see!' Unsatisfactory. Frustrating. Very Robert.

The first of my questions was answered as we turned onto the Southwest Highway: the property was to the south of the city and obviously not by the sea. As we travelled deeper into the country, the Darling Ranges gradually started to unfold to the east of us. Flocks of twenty-eights and black parrots rose in clouds from the roadside as we sped by and kangaroos forced by the lack of rain to graze on the lower slopes raised their heads to stare at us in a singularly incurious manner.

After about an hour's drive, we turned off the highway onto a dusty, rather corrugated track and just a little way down this road, Robert stopped. This was Boyd Road. This was the entrance to the farm. We had arrived.

How quickly excitement can turn to apprehension. It didn't look promising at all. I happen to hate barbed wire and the sagging fence made of this particularly unfriendly material was the first thing I saw. Robert was getting out to push at a rusty gate that was all but off its hinges. Beyond that, I could only see bare ground or dry grass, cut here and there by deep tracks. I felt as depressed as I had been excited a few moments earlier. Where had all that adrenalin gone? My imaginings really had let me down. I was terribly disappointed, but to be fair I could hardly blame Robert. He had certainly gone out of his way to say nothing. At the same time I had been totally unprepared for the neglect that lay ahead. Driving rather carefully onto the property, the large car only just managed to squeeze between a huge boulder and a rusted wheelbarrow lying on its side. Farm implements, apparently discarded wherever they happened to break down, lay half-hidden in the long dry grass. There was rubbish—old car tyres, piles of wood, brick rubble, a greening copper—as far as the eye could see.

Well, this was the farm. I searched for a positive aspect, but everything looked hot, dry and rather hopeless, as if it had long ago been given up on. I couldn't help the comparison with my childhood farm which only made me more miserable. And then I felt guilty at my feelings of disappointment, which I suppose were more a question of frustrated expectations. For Robert's sake, I thought I had to make an attempt at equanimity or at least pretend to be noncommittal. I raised my chin and looked straight ahead.

But it was to get worse. As I went to get out of the car I spied what I suspected was the farmhouse. Over the years, what had been a rather dull and ordinary building to begin with had been patched many times over with odd lengths of corrugated iron, scraps of canvas and whatever other bin ends had been to hand. It resembled a poor-white dwelling more than anything I had seen in this country so far and was more in need of demolition than repair. I gripped the leather seat rather fiercely, searching for something to say. It is very rare that I have to search for words, but all I could find to comment on were the flame trees, one of which was in full bloom. By now Robert had guessed at my disillusionment. It was a small gesture that didn't quite come off, but I knew he was trying to reassure me when he said the house would be one of the first things to go.

I looked away but as I did so my gaze was held by a tall rose bush. It was sitting quite by itself in a patch of parched soil just a little way from the house and from the condition of its straggling branches and yellowed hips it was as much unloved as it was unpruned. But despite this lack of apparent care, the bush was aflame with a flush of the most glorious red roses, a show made all the more remarkable by its surroundings. I wondered how it had survived, let alone continued to bloom so vibrantly. As I watched, a breath of breeze tugged at a stem and one of the roses released a drift of petals which settled nearby on the ground. For a split-second I was back at Chobe watching flashes of the same shade of scarlet as the carmine bee eaters darted for insects along the banks of the river.

Meanwhile the farmer had come over. He shook Robert's hand, nodded at me and suggested he drive us around to 'explain a few things'. All three of us squashed into the front seat of his ancient Land Rover, and soon we were being jerked rather violently over the rutted tracks that were used to get from one side to the other of this so-called farm.

I was still doing my best to prevent my face from giving too much away but I could sense Robert glancing at me from time to time. The new property was supposed to be a great surprise for me. It had indeed been a surprise, although perhaps not quite in the way Robert and Janet had anticipated. It was hard to believe that only a few short hours earlier I had been worried that I wouldn't be included in the move. Compared to this, my Perth flat was a slice of heaven. But Robert and Janet had felt the farm had potential on first seeing it, and had purchased the property on that hunch.

I could see that Robert certainly had no regrets. His gaze swung from side to side as we travelled up through the hills and when we stopped at a point far above the lower slopes of the farm, he was the first to swing his long body out of the vehicle. Shading his eyes against the Western sun, he looked down on what could only be described as a thoroughly unprepossessing property. And yet he didn't seem to see this. He spread his arms. 'One day all this will become a *fabulous* thoroughbred horse stud,' he predicted. I said nothing. What could I say?

Time was to prove him right, but of course I didn't know that then. The word 'fabulous' teased my memory. I had heard him say it once before. But when?

Almost completely ignoring the presence of our host, whose driving skills were well challenged by the track, Robert continued in a chatty mood. He reminded me I had created a very successful riding school in Rhodesia soon after World War II. 'Don't you remember that you started with only a few acres of bare land with a ghastly old building? And that it wasn't long before you tore that down and built the stables?' He talked quietly and deliberately, bringing up the past, encouraging me to draw my own parallels.

He reminded me that when my late husband Charles and I took on the task of creating a tourist camp in a remote area of the Bechuanaland Protectorate, he had been very uncertain of its future. The chances of success in creating such a sophisticated tourist resort in the middle of the African bush were not in our favour, he had pointed out very clearly. But by that time we had gone too far to stop and, besides, we had faith in our own abilities. While there was no doubt that we certainly had our share of problems, there was nothing that could not be overcome. By the time we had finished, the luxury Chobe River Hotel stood grandly alongside the original tourist camp on the banks of the beautiful Chobe River. It was set just outside the Chobe Game Park, later to become a National Park, home to some of the richest wild animal and bird life in the world. We had achieved what we set out to achieve. My memory lurched. I remembered now what I had been trying to bring to mind earlier: a very much younger Robert standing in the grounds of the near-complete hotel gazing out across the river. The sun was starting to sink and the elephants were just arriving at the far bank for their evening drink. He had turned to me. 'You were right, I was wrong,' he had admitted then. 'It really is *fabulous!*'

The climax of this first drive was bumping along the tops of the ranges along badly pot-holed tracks into which each wheel seemed to drop alternately, each time throwing us hard up against each other. Or sometimes a sharp turn off the path would offer a smoother ride across the crackling grass. Trees were definitely more plentiful up here: tall red gums and other eucalypts unknown to me with silken silvery trunks. The view stretched before us like a Disney panorama, sweeping down over the foothills with their rocky outcrops, and out across the flats of the coastal plains to the ocean itself. In many ways it reminded me of Africa where I had driven hundreds of miles over similar terrain. The difference here was the grass trees, eaten almost to the stumps by kangaroos. Our host interrupted Robert's monologue. 'Two years ago, we had a bush fire up here. Fright'ning thing it was. After that, them black boys was just stumps.' They were still stumps in most cases, gnarled, blackened, just starting to shoot from their centres. I imagined one rather intricately woven trunk as a young elephant. Another reminded me of a reclining cheetah. This was something I had been missing so very much in Australia: my beloved wildlife. It worried me a little that I was creating animals out of shadows.

In another burst of conversation, the farmer told us this property shared a boundary with the State Forest, that it was the home of many kangaroos and beautiful birds, as well as possums and other small creatures that are not always easy to find. I realised that Robert, well aware of my love of the wild, had chosen this farm both to create his own dream and to keep his promise to me. It was at this point that my outlook started to change. To be fair, I thought, this was hardly the time to view a West Australian property at its best. We were at the end of the dry season with the temperamental rains due anytime now. The grass was so sparse that even the few cattle we passed were pulling hay from bales.

The drive home was very different from the outward journey. Robert was still inclined to talk and talk he did. There was no doubt that what he had in mind for the old farm was a vision and that to put it to rights would take an enormous amount of work—and money—but so persuasive was this rather rare burst of enthusiasm that I was starting to believe it might indeed be an achievable goal. Perhaps I wanted to be persuaded. Certainly my life had been a series of challenges. Why should I think that they would stop now? My negativity began to dissipate as I started to think the project through properly. The long silences that were part of Robert's cogitating processes were punctuated with short sentences as he outlined his plans. A few repairs would be done to the old house which would be kept for as little time as possible, he said, and a manager would be employed to start clearing the rubbish and ripping out the barbed-wire fences. As for horses...as soon as there was a suitably fenced paddock and shelter, he suggested we begin by collecting Impion, his polo pony, from where he was stabled at a small riding school and transport him to the farm. This meant that instead of exercising him only at weekends—the highlight of my week—I would be able to ride whenever I wished. I started to feel better and better.

But there was more. He also suggested I source another polo pony so we could ride together around the farm, planning roads, paddocks, stables and accommodation. For a moment he took his gaze off the road. Quietly he repeated his earlier conviction. 'Ethnée,' he said, 'There is absolutely no doubt in my mind that this will become the best thoroughbred nursery in the southern hemisphere.'

That night I hardly slept, wondering where we would begin. I shuddered to think of all the barbed wire that would have to come down and of the cost involved in replacing it. Then an image of the glorious rosebush with the carmine-coloured roses popped into my mind. I wondered who had planted it and whether that person had been around to watch it bloom.

Barbed wire and roses fought for precedence in a series of troubled dreams. But morning brought the clarity that had been missing the night before: Who or what could stop Robert once he had set his heart on something? He was already committed. Robert's promise was about to become a reality.

CHAPTER THREE

The Starter's Light

As it happened, the couple who owned the riding school where Impion was stabled were at the point where they were particularly keen to sell the business and move on. Their suggestion that they take over the management of the farm came at just the right time. But where would they stay? I had been hesitant in mentioning the accommodation. I still regarded the old house as a monstrosity. It rambled, each room radiating out from the living room almost as if each after the first had been an afterthought. The sleep-out looked positively dangerous. This appeared to have been tacked arbitrarily onto the last room built, walled up with badly matching pieces of weatherboard and the lot topped with a molding canvas awning. It looked as though a gentle tap would push it over. The whole building, such as it was, badly needed painting. Probably the best thing about it was a rather large fireplace in what must have originally been the living room and which in fact still drew extremely well.

'As a temporary measure only, do you think you could find it livable?' I had asked our prospective managers rather tentatively once they had had a chance to see it. I was relieved when they made fun of my misgivings. They would make the best of it for the time being, they said, and in no time they had moved in and started work.

Robert kept his word, however, and it was not long before alternative accommodation was built and the old house demolished. After the bulldozer had finished and the dust had cleared, all that was left were the trees and, of course, the rose bush. The post-and-rail paddock that was eventually erected in this area we christened Rosebush Paddock in its honour. That's what we continue to call that paddock to this day.

Meanwhile, although I was still living and working in Perth, I had started to spend my weekends at the farm in a borrowed caravan that I had parked under a stand of enormous red gums for shade. By this time the reservations that I'd initially had about the farm had largely disappeared. Certainly there were difficulties, but I loved every moment of once again being on a property with horses. Perhaps the only disadvantage in my living conditions that caused me any bother at all were the wretched honky nuts that bounced day and night like gun fire off the steel of the caravan roof.

But some time before the old house was taken down, in fact only six weeks after I had first seen the farm, there was a great deal of excitement as Robert and Janet's weekender was trucked onto the property. This was a Bunnings transportable cottage painted in an unrelieved—and unrelieving—mission brown. No one appeared to have any clear idea where it should go, but after some lengthy deliberation and a lot of head-scratching on the part of the openly bored truck driver, it was decided that the driest, flattest piece of land lay just to the south of the existing house. This was where it was finally off-loaded to the little disguised relief of the team of three men who lost no time in taking off for their Friday night tipple.

As it happened, my caravan was within sight of this cottage, so I was there that weekend when it started to rain. Down it came, starting late Friday night, and continuing just as heavily all of Saturday and into Sunday morning. These were the rains we had been waiting for...doubled for good measure. The dry land around the cottage turned to mud, the mud turned to sloshy water, and the brown building rapidly became indistinguishable from its surroundings as either it began to sink or the water to rise.

When Robert arrived on Sunday around lunchtime it was obvious we had picked the wettest part of the whole property on which to site his weekender. And equally obvious, from his creased brow and raised eyebrows as he scanned the waterlogged land, that he wondered why. We avoided his gaze and concentrated instead on the logistics of deciding where to re-site the building.

In retrospect, of course, the rain was timely. At least we knew where not to put the cottage; furthermore the next site would be Robert's responsibility. On Monday morning, the reluctant three were recalled, the weekender was winched onto the trailer, the tractor hitched and our little group ordered to stand clear of the mud spray. But even this wasn't as easy as we would have hoped. The engine revved, wheels spun, but the mud wasn't going to give up quite so readily. When finally the weekender did break free, it did so with a gurgle and a rather obscene sucking sound which brought a round of hearty cheers and a 'you beaut, mate' from the men. When next offloaded, it was onto what was to become its permanent site about two hundred metres from where it had first been put down. From the windows to the east you looked out to where the ranges met the sky. On the other side, the sun streamed through the western windows and it seemed as if it were only the trees that stood in the way of an ocean view.

It had made sense to place the new cottage as close as possible to the site of the original house since the all-important electricity pole was close at hand, so light and heating were easily taken care of. Water was more difficult. This was long before the advent of scheme water and even before the first dam was built. For quite a while, the only water supply for the cottage was a 44-gallon drum which sat outside the back door. As the rainy weekend had already hinted, the site we had first chosen for the cottage turned out to be one of the wettest on the whole property. This would become the site of our first dam.

Although in its first year the cottage was only used on weekends by Robert, Janet and their two young children, in a comparatively short time it resembled the hub of a wagon wheel. As the development of the stud raced ahead, the cottage would prove to be its nucleus. One spoke would lead to the workshop, another to the first stable block, a third to the first post-and-rail paddocks. Still radiating out from the cottage and next to be built would be the veterinary area and serving barn, followed by the office block and then more staff quarters.

I can't quite remember when we stopped referring to the property as 'the farm', but as it became more and more focused on horses, it seemed less and less like a farm. The stud had started to evolve.

Robert wanted me to fly east to New Zealand and South Australia to source the brood mares that were to form the first of our breeding stock. Although I was extremely proud to think that he had so much confidence in my judgement, I had already made plans to meet my younger son Simon for a sailing holiday. We had decided I would spend a month with him aboard his yacht *Maggie May* and this new development presented me with a difficult choice. I couldn't be in two places at once. Then I thought: why couldn't I? Why not combine both trips? And that is what I did. The sailing holiday in Fiji would be followed by a business trip to New Zealand.

Simon looked tanned and happy. It was two years since I had last seen him in Africa. When Bechuanaland became Botswana on Independence in 1966, he had decided to give up the job he'd held for many years as game warden and take advantage of the handshake and pension he was offered. Since then he had been fulfilling his own dream. He had bought the 30-foot yacht in 1969 and was now sailing around the world, stopping off at some of the lesser known almost completely unspoiled islands—of which the Galápagos off the west coast of South America were his favourite—making a documentary of marine life in the process. I was captivated by the stories he told in the long warm evenings as we sat, mellow with sun and wine, on the sloping deck of the *Maggie May*. He spoke of the route he planned to take, of the places he intended to visit. He told tales of island ceremonies strange to me, of an island currency made from the plumage of red sunbirds, of rare birds and animals, sacred sharks and miniature dragons. I listened mesmerised for as long as he could be persuaded to reminisce.

One evening he was unusually quiet and I took the opportunity to ask him a question that had been on my mind. 'What advice would you give others who feel drawn to adventure, Simon? Obviously there are hardships and compromises that you have to make in order to do all this. What keeps you going? What makes it so worthwhile? Is it the adventure itself that is so attractive or the knowledge you gain along the way?'

'The freedom', he said without hesitating. 'The freedom is the most important thing. Only secondly, the adventure. For anyone dreaming of similar adventures, I can only give one piece of advice. Don't delay. Put your dreams into practice. The first step is always the hardest because doubts and fears set in…but the eventual rewards make it all very much more than worthwhile. Never let fear hold you back.'

Doubts and fears. As he said that, I was reminded of my feelings on first seeing the farm. I had certainly had doubts and fears aplenty. Although I have met and overcome many challenges in my life, I wondered then whether I would have gone ahead with the stud if it had been my decision alone, if it hadn't been for Robert.

My two sons…as different as the sun and the moon. Robert was the money-maker, Simon the dreamer. Some years later when Simon's documentary was complete, Robert offered to handle the deal and promised that in doing so, he would make a million for Simon. Simon's rhetorical 'What do I want a million for?' says it all. The documentary was later bought by Australia's Channel Nine.

And so on to New Zealand. Another country, another contrast. I was met at the airport in Auckland by an agent who whisked me off to numerous studs. He was certainly both helpful and charming and there is no doubt that New Zealand breeds some champion horses, but on the whole the trip was rather disappointing. In the interests of keeping their leading bloodlines to themselves, no one was prepared to sell us really good mares. I must say I learned a lot on that trip.

I reported this to Robert during our next phonecall, but as usual he was one step ahead of me. He interrupted my explanations. 'Don't worry about that,' he said. 'I've arranged for an agent to show you around some of the South Australian studs. There's a flight to Adelaide tomorrow morning and I'll get him to meet you off the plane. I have a hunch that there are several mares there that might just suit us.'

I was met at the airport in Adelaide by a man called David Coles who proved to be another marvellous escort. Again I visited a number of different studs, some of them very beautiful, others both aesthetically beautiful and successful like Lindsay Park Stud with which we were to form an excellent business relationship in time to come. But it wasn't until we arrived at the last stud on our agenda that I saw what I really wanted. Here were six mares, all of which looked as though they might be suitable. I photographed each from several different angles, collected their pedigrees and flew back to report once again to Robert.

The photographs were not enough; it was typical of Robert that he wanted to see the mares for himself after all. He flew over to Adelaide. But how thrilled I was when he called me a short time later to tell me he had decided to purchase them all. On the same trip, he also bought six yearlings that were to be readied here and sold at our annual yearling sales.

Now that we had our first crop of broodmares, it was time to officially name the stud. All along there had been no question that it would be anything other than Heytesbury Stud. Heytesbury was the name of the family estate in England where the Holmes à Courts had lived for nearly two hundred years. Although the family had sold the property with its glorious Elizabethan mansion and beautiful parklands well before World War II, the name Heytesbury was to become an obvious choice for all our homes. Our house in Rhodesia, now Zimbabwe, had been named Heytesbury, and now Heytesbury lived again as a stud with a golden future.

All the stud needed to make it complete was a top stallion. Robert was adamant he wanted only the best, preferably a past winner of Australia's premier race, the Melbourne Cup. I contacted an agent I knew, Bob Maumill, relayed what Robert had in mind and asked him if he knew of any likely possibilities.

'I just might,' he said. As it turned out he wasted no time in ringing Sir Walter Norwood in New Zealand, owner of the stallion Silver Knight, recorded as winning the Melbourne Cup in 1971 in the second fastest time ever.

Bob called back with the information that the stallion was for sale…for a price of $65,000. Robert's decision was instant. He replied immediately: 'Where do I send the cheque?' There could not have been a faster or a better deal.

Heytesbury Stud was off and running.

CHAPTER FOUR

A Handful of Southerly Wind

But it was in that frenetically paced period of a year or more between the choice of the first crop of broodmares and the arrival of Silver Knight that the physical shaping of the stud took place. Although I was still residing and working in Perth during the week, I lived for weekends. The rest of the world had ceased to exist. My life began and ended at the gateway to the stud.

I lost count of the times Robert, Janet and I sat down together to discuss the developing stud. If it were to become the success that we were determined it would be, the stud needed a framework, a design. Robert and I would often ride our horses up into the hills to view the property from the scarp. From there we could visualise what the layout should be; we could see exactly where we would site the yards, the shelters, and the larger paddocks that would follow.

The corroding barbed wire was removed and dams, roads, fences and trees took its place. Gradually the property evolved, sometimes with a maddening slowness... sometimes so quickly that changes seemed to happen overnight. An infrastructure began to emerge.

For Robert, dams were the priority. Horses need a lot of water and without water, the stud simply could not operate. The first dam had already been constructed in the boggy area of the weekender's first landing. Next, the only spring-fed dam on the property, at the top of the hill, was deepened and enlarged. But we needed others and for months the air was thick with the noise of revving bobcats hollowing out cavities and strengthening banks before the graders and tractors took over to landscape ready for planting.

Still this wasn't enough. 'Now I want a big dam,' Robert stated. And so a large dam was built to take advantage of a stream running down a hill. That first year we had an excellent rainy season and the dam soon filled. We watched gleefully, marking off the measuring posts as the water rose steadily to the 12-metre mark. Since both the dam and the walls surrounding it were new, we installed a sluice gate to allow any excess water a harmless escape route. Although the water in the big dam never again reached the quite remarkable height of that first season, the dam is electronically equipped to feed several smaller dams. These in turn irrigate thirty-one special yards planted out with kikuyu and clover, which means the mares and foals always have green pasture. Less important to the business of the stud, but a constant delight to us, are the hundreds of wild duck that make their home in the vicinity of the dams, some of which occasionally come down to graze beside the horses. One year, three black swans stopped by.

Of almost equal significance to the dams were roads. There were no original roads, merely a number of half-hidden tracks which meandered from one part of the property to another. Meandering wasn't for Heytesbury. And on the flat there was nothing to stop us building roads that led logically from one area to another. Graders came and went and the dust seemed to hang in the air throughout that second summer until, finally, the gravel was laid.

Almost simultaneously with the building of the roads came the tree planting. Citidora saplings were delivered, literally by the thousand. In fact, the first batch consisted of some four thousand trees. What Janet did not plant herself, she organized to be planted, so we all ended up becoming involved... visitors, too. Although the young trees were no more than a foot high at first, it was not long before they started to appear first along the existing ways and then down each side of the new roads. Right from the start, and small as they were, the trees defined the property and this glimpse of the tree-lined avenues that would one day be a trademark of the property allowed us to feel we were getting somewhere.

Through all these massive planting projects, Janet had one aim that dominated all others: the indigenous nature of the property was to be kept sacred. There were few plantings that did not seek to protect the natural look of Heytesbury, and it was this combination of effort and foresight that not only made the property so definitely Western Australian, but which also stamped it with the feeling of authenticity that remains one of its leading charms. The lawns, verges, avenues and shrubs were all designed to fit in with the original grand old red gums and grass trees. The road that passed alongside the cottage was aligned to run around the tallest grass tree I have ever seen. It must reach more than thirty feet into the air and it's said to be more than 200 years old.

Janet's plantings weren't restricted to eucalypts. Fragile-looking paperbarks were planted in groves of three or four alongside some of the dams, an avenue of flame trees was planted to the east of the cottage and around the area that would become the foaling-down yards. And of course the few stands of existing trees were carefully protected. Beyond the lower paddocks is a natural forested enclave where native banksias, Dryandras and Casuarinas all flourish. High up in the hills, on the steeper slopes, the terrain is almost totally dominated by enormous jarrah trees, which appear to cling so precariously to the gravelly slopes that it is a miracle they have survived both fire and the stresses of our unforgiving hot, dry summers.

Robert's vision was becoming a reality. Although each phase of development inevitably carried its own welter of decision making and problem solving, as each obstacle was overcome our sense of pride in the property grew. But our pleasure in our surroundings was secondary. Of prime importance was that every move we made was enacted with the business of the stud in mind, to which the well-being of the horses was, of course, central. Quality feed, abundant fresh water, adequate shelter, and paddocks large enough to allow space to work off high spirits were our aim. Horses seem to prefer a sheltered outdoor habitat to manmade cover. They really thrive in natural surroundings if they can find refuge from wind or sun whenever they need it, which is more often than not in this part of Western Australia. Consequently the areas we fenced off were those that offered part sun, part shade. All the paddocks were spacious and in some we planted a spinney—a stand of trees with an undercover of smaller shrubs protected by a post-and-rail fence—to act as a windbreak.

If all this sounds easy, perhaps I've only mentioned the favourable outcomes of those early days. Because far from everything we attempted was an instant success and it was more often than not a case of trial and error in learning to live in line with nature. Sadly, even the jarrah trees aren't immune to some of nature's tricks. For half a century now, an imported disease has been causing dieback, a condition that ultimately kills the tree from the top down, starting with the uppermost leaves and branches. In the West, while all other eucalypts are immune, the jarrah suffers heavily and experts have as yet been unable to find a solution. It is a particularly grave problem in the forests of the Darling Ranges.

The horses, too, are not above mischief. Planting the trees was one task, but ensuring that the horses didn't chew on the trunks—which they love to do, and which causes almost instant death to the trees from ring barking—was another. This meant we had the additional chore of encasing the trunk of each growing tree with protective six-foot mesh. You can imagine the scale of this task—there are a great number of trees. I'm reminded of the story of a young veterinary student who came to us for work experience. He was warned by a member of staff that he could do most things on the property, but that he must 'never damage a bloody tree!'

And we had to learn to live with the winds. The first avenue in front of the cottage, for instance, we planted with River Red Gums. But we had no idea just how savagely the easterly winds can blow. They pour over the tops of the scarp and swoop down onto Heytesbury showering cottage and horses alike with dust and pebbles, and generally causing amazing havoc. Since the River Red Gum Avenue runs from north to south, the prevailing winds caused all the trees along that stretch to lean to the west. Although over time they grew into, rather than out of, their hump, becoming quite attractive, we learnt our lesson and after that, as much as possible, we planted east to west.

We found that settling in the lee of the ranges also left us particularly vulnerable to the aftermath of cyclones. The tail-end of Cyclone Alby was particularly vicious. It whipped back and forth across the flats, causing a considerable amount of damage. Some of our new trees were so badly affected they looked as though a fire had been through them. The grass also looked as though it had been blasted with a blowtorch. An old tin shed was dismantled, huge pieces of tin arching through the air and landing several paddocks away, only to take off again and again, resting between gusts before rolling onwards like saltbush caught in an eddy.

But of all the events of those early years, the one that lives most warmly and vividly in my mind is the arrival of Heytesbury's first foal. The first crop of mares had settled in quickly and happily. Pending the arrival of our own resident stallion, Silver Knight, the mares had all been served outside Heytesbury the year before and were in foal to various stallions.

It looked as though a little black mare called Tixall would make history by producing the stud's first foal. I felt particularly close to Tixall as she was one of the foundation mares I had chosen personally from the South Australian stud.

Is it a contradiction that I was both extremely excited and just a little apprehensive at the same time? Although I had been involved with horses all my life, and although I had also done a fair amount of studying in and around the various aspects of studwork, which included attending a number of university lectures on the subject of foaling, this was a first so far as the practical side of studwork was concerned. I had certainly never delivered a foal.

But nothing could have kept me away and it was I who elected to keep watch through the night. Robert and Janet were down for the weekend and had spent some time watching the mare until, as midnight approached, they decided she would not be ready to foal down for some time and went off to bed. The manager, too, had retired early asking me to call him when the birth became imminent. Since he lived next to the paddock, help was not far away. Or so I thought. At this point, fortunately for my peace of mind, I was not aware he was as new to the practice of foaling as I.

I planned to spend the night in my car, a Rover 2000 that Robert had given me as a birthday present. It was parked close enough to the paddock to observe the mare, but not so close as to be intrusive. My Rhodesian Ridgeback pup, Rachael—the first ridgeback I had owned in Australia—was curled up on the back seat of the car with her blanket and hot water bottle. I pushed the front seat back to a more comfortable reclining position and we settled down to wait.

❖

I am suddenly aware that the car windows are steamy and that, outside, the mare is becoming restless. She starts walking. I step out into the cold dark of the October night to feel that the wind has set in already. The skinny beam of my torch shows Tixall's dark body made darker still with sweat. Her tail is held high and she is preparing to lie down. Moments later, apparently unaware of me—and of Rachael by my side—she goes down. I am shaking. There is no moon and, although the sky is awash with silvery stars, so little light. A more powerful torch would be a godsend. The mare's contractions are becoming stronger. I check my watch. It is just after 2 a.m. Time to call the manager. I bang on his door until he rushes out with another torch and together we witness the first birth at Heytesbury Stud.

One little hoof appears, and almost simultaneously a second comes into view followed by the foal's nose; for just a moment nose and hoofs remain suspended, tucked one above the other. The mare groans her way through a last contraction and a few seconds later the foal slips out in a rush, to be greeted with a delighted whinny and a big lick from Tixall. The umbilical cord ruptures and breaks free, and the slightly dazed-looking foal lies on the ground beside its mother. Robert and Janet come up. I look at my watch. It has been twenty minutes from start to finish.

Shortly afterwards Tixall heaves herself to her feet and all four of us are forced to swallow hard as we stand, inadequate torches lowered, watching the unashamed display of mother love: the mare standing over her young black colt, nudging gently, encouraging him in his first efforts to stand and suckle.

Nothing in my life so far had prepared me for the experience of watching the birth of a foal. And even though I have now been present at literally hundreds of births, assisting with some, observing others, no matter how often I am privy to this event it is always indescribably special. I never fail to admire the brood mares and the depth of love and care they show their offspring. Just occasionally a mare may reject her foal. I witnessed this in Kentucky years later, when a horse refused to go anywhere near her foal which was quickly adopted by a dear old Percheron mare.

Tixall's colt would not be abandoned. The morning star was blinking brightly and the sky was just starting to turn a pale pink. The foal had progressed from his first shaky steps to tentative and somewhat clumsy cavorting about the paddock. It had certainly been a magical night.

We retired to the cottage to allow mother and foal time to bond with each other and Robert opened the bottle of champagne he had been saving for the occasion. As we raised our glasses to Tixall and her foal, he said with a satisfied smile: 'And now we have really started Heytesbury Stud. We'll aim to breed a Melbourne Cup winner'.

I quoted a little saying I had come across:

'And God took a handful of southerly wind, blew his breath over it, and created the horse.'

CHAPTER FIVE

A Knight to Remember

It was ironic, although I'm not sure it occurred to me at that time, that although we had been unable to bypass New Zealand's not unnatural refusal to part with any of its top broodmares, a short while later we purchased one of the country's prize stallions instead.

Silver Knight (NZ) by Alcimedes out of Cuban Fox was foaled at Trelawney Stud in New Zealand in 1967. He was branded with the figures 3 over 7 and he certainly brought a lot of luck with him when he crossed the Tasman to his new home in Australia. Over the years I have had all my personal horses branded with either a 7 or a combination of figures which will add up to seven. Just as the Chinese believe in the positive powers of this number, so do I. It has always brought me luck.

Some horses react badly to flying and Silver Knight was one. Even though the flight from Wellington to Sydney only takes a few hours, it was thought best that he take the sea route travelling in luxury on the trans-Tasman ferry. His strapper and devoted companion for the trip was a young woman who worked at the stables of the stallion's New Zealand trainer, Eric Temperton. She was a tiny girl called Janet Thom who had worked closely with Silver Knight for years and she was terribly distressed when she had to say goodbye to him.

Robert had sent another strapper, Ian Roy, across to Sydney to take over from Janet and Ian caught the brunt of her distress. He did his best to calm her and promised we would write to let her know how he settled into his new home. As it turned out, we wrote to each other for some time.

Ian's task was to travel overland with the stallion to Heytesbury. 'Take your time. Stop wherever and whenever you feel you need to,' Robert advised. We were determined the lead-up to Silver Knight's arrival would be as enjoyable as possible. So his first encounter with his new country was not just a boring float ride. Instead he was unloaded frequently to uncramp his legs in the parking bays or to graze on the long grass at the edges of camping grounds.

Wherever he went Silver Knight attracted attention. Like many greys, he had been born black, but as a four-year-old his coat was a glossy steel grey with a silver mane and tail. Perhaps people were surprised to find such a sight after stretches of gloomy grey bitumen, but Ian returned with tales of cars suddenly sweeping off the road, their passengers stopping to chat and admire. They were thrilled to learn that the stallion was a Melbourne Cup winner travelling to his new home in the West. The time was right for Western Australia to take its place in world-class racing.

After such a gentle and loving start, no wonder the stallion adjusted so well to his new life. After an inspection of his surroundings, including a brand new brick stable we had built for him and a large paddock with huge shady trees overlooking our first dam, he settled in to live grandly as befitted a horse of his calibre. His name and an impressive pedigree that covered several generations were painted on to a metal board and nailed to his stable wall for all to see. He was a true knight, a gentle and intelligent horse, kindly disposed towards the wild duck that visited his paddock each evening hoping, rather in vain I fancy, for grains of leftover feed, and our tame guinea fowl and peacocks which could often be seen in his company. Our stallion man took him for regular walks along the newly tree-lined avenues or up into the ranges and Robert would ride him whenever he came to the stud.

Shortly after his arrival, in his honour and somewhat sentimentally I suppose, we planted what is now a magnificent avenue of lemon-scented eucalypts with splendid silver trunks along the track leading to his paddock. The scent is almost overpowering, particularly in the early mornings as the dew evaporates from the leaves under the warmth of the sun.

When the day arrived to float Silver Knight to Belmont Racecourse to parade and take bookings for his first stud season, the manager decided it would add to his glory to rinse his mane and tail with a substance called Magic Silver White, widely used in that era for adding interesting streaks to ladies' hair! Unfortunately, use too much and it turns out a disaster, as many a schoolgirl knows only too well. In this instance, Silver Knight ended up with a bright mauve mane and tail and all the washing in the world failed to wash it out completely in time for his first showing. But after an exhibition gallop, mauve tail notwithstanding, little murmurs of 'most impressive' could be heard from among the crowd. From the proud and eager performance he gave, I wonder whether he expected to win another cup that day? Whatever he had in his mind, he certainly showed himself very well and filled his stallion book for the coming season. Now he had a different life from racing. How many champions would he sire? Would one of these be Heytesbury's first Melbourne Cup winner? Or, more immediately, which of our six foundation mares would be the first we put him to?

When Robert called me for a meeting to discuss the first stud breeding programme, I wasn't to know that it would become something of a tradition between us—sitting down over the stud books every year sometimes talking for hours into the night or early morning. Perhaps because he didn't drink himself, it did not occur to him to offer me a glass of wine, even a cup of coffee or a snack. We would just talk and talk, filling in the books as we went. It was something I missed terribly when he died.

Typically, on this evening, he had already decided which mare we would put to Silver Knight first. He seemed to have a sixth sense as to what would produce the best results and for this mating he suggested Brenta, a pretty brown mare with a lovely nature. She had attractive white markings: a star and thin stripe, and a small white sock on each of her hind pasterns. More importantly, she was a sprinter while Silver Knight did better over longer distances. Although she had not had a particularly impressive racing career, Brenta had been purchased because her breeding by distinguished sire Coeur Volant appealed to Robert. She had cost only $600.

Robert was talking to me—comparing and contrasting the two horses' conformation and colouring, their track records and pedigrees—but I could see his mind working behind the words. He stopped suddenly and smiled, his pen hovering over the page, 'It'll be a good match. I know it will.'

He turned out to be right. It was destined to be a most propitious match. Silver Knight appeared to be delighted with our choice, and I think Brenta rather fancied him too. But on this first occasion, Silver Knight's performance left us a little stunned. One moment he was enthusiastically carrying out his duties, the next—much to our consternation and Brenta's surprise—he had fallen off the mare and collapsed onto his side, quite unconscious. To our relief, it was not long before he revived. As it has turned out, that is the only time one of our stallions has fainted on the job.

Later on in the season, as he became more used to his new career, Silver Knight became rather fussy. When he was taken into the breeding barn to meet a mare he showed a great deal more interest if she happened to be a chestnut, and this preference for 'redheads' only grew stronger as the years passed. One morning we had a tremendous amount of trouble getting him to take any interest at all in the rather old and admittedly somewhat unattractive plain bay mare that awaited him. She had no pretty markings, no red coat. He was totally disinterested. In the end he had to be taken for a walk during which his responsibilities were outlined very clearly. When eventually he was brought back to the barn it seemed he understood, because he decided to undertake his duties, if reluctantly.

Robert had an unusual plan for Silver Knight. In order to ensure that the horse did not become bored or gross during the break from stud duties, he decided to give him a complete change of environment. So after the first season, he was put back into light work at the stables of trainer Ted McAuliffe, near Ascot Racecourse. What a change from pacing back and forth along the fence, eyeing the broodmares, searching for a chestnut and a little flirtation. He must have thought his racing career was about to start again. Each morning he was taken to the track for an easy work out. In the afternoons, he was led out for a walk and perhaps a picking of green grass. He hated swimming, so that was not on his itinerary. The first and only time he was taken to the pool, he sank; he just went straight down to the bottom while we looked on in horror. But apart from this experiment, the stallion thrived.

Although only 18 months had passed, we had achieved a great deal since we first drove onto the property in 1971. The first managers had lasted a little over a year. We then employed somebody else but it did not work out quite as it should.

It was at this point that Robert dealt me another surprise. 'Why don't you move down to the stud fulltime and run it. You could live in the weekender for the time being and I'll build something else for us. You know more about what's going on and what should be done than anyone else. What do you think?'

This time Robert was the one who had to wait for me to speak. Rather impatiently, he repeated, 'Well, what do you think?'

I was unsure as to what I thought. Run Heytesbury? Keeping a watch on the stud was one thing, but running it was quite another. And what exactly did he expect me to do? What were my duties, responsibilities? What about my life in the city?

My mind went careering over the problems, the possibilities. Although I adore horses and had owned a successful riding school in Rhodesia where I taught everything I knew about dressage, cross-country riding and show jumping, I had an idea that the skills required in this new enterprise would be altogether different. This, after all, was a stud and we aimed to breed top racehorses. Even the grazing was different from that in Rhodesia, and inevitably the feeding programme would differ.

I told Robert that I needed time to make up my mind. I sensed that he already knew I would take it on, but I needed an interlude to think and to find out exactly what was involved.

Over the next few weeks, I set out to learn as much as possible about the stud business. I spent some time with a feed merchant who was very helpful in discussing that vital aspect of horse well-being with me. I visited other studs across the country including the few that were in business at that time in Western Australia. I attended whatever courses and lectures there were on offer on horse management and foaling at the University of Western Australia. My head was buzzing with the new information I managed to collect, but I was still not certain. What to do? It was so unlike me to prevaricate. Perhaps a week living fulltime at the stud would make up my mind for me one way or the other. I called Robert and told him that I would give it 'a trial run' and surprised myself by the firmness in my voice when I added that for once I would not be hurried.

I moved down to Heytesbury in the January of 1973. It was one of the many days I was to experience when the mid-summer heat trembled in a haze over the stud, challenging the new irrigation system and trees still too young to be of any use for shade. Inside, the weekender was airless; outdoors, the grass was a crisp, baking gold. It felt as though the ranges towering above the stud had sucked all the air out of the day and the sea breeze had not yet arrived. It was very still. All my furniture, my pictures, books and mementos from Rhodesia lay in cartons on the living room floor. I looked at my dog Rachael—and she gazed back at me. Then she whined and ran to the open door. Her ears pricked and her brow creased. She looked back at me and gave a small bark. 'Come on,' she seemed to say. 'Opportunities for endless walks without a lead lie out there.'

Rachael and I were gone for some time. A real reconnaissance of the lower reaches of the ranges that I rarely had time for on my busy weekend trips. Once we got back, we came back home. The weekender had become my cottage. There was no decision to make. I started to unpack.

CHAPTER SIX

A Slow Track

Now that I was down at Heytesbury fulltime, I was closer than ever to the constant and inevitable demands of the developing stud. There was something happening all day and every day, not only from dawn to dusk, but throughout the night as well. It was often difficult to disentangle the normal stream of stud proceedings from unexpected, and sometimes downright painful, misadventures.

Looking back down the years, once we had established a routine, as predictable today as it was then is the seasonal aspect of stud happenings. These days we just do it a lot more scientifically and in some ways, we are smarter. The year starts from the horses' birthday—which is on August 1 in the southern hemisphere—and which heralds the beginning of the busy foaling-down season which can go through to Christmas. Almost concurrent with this is the stud season, starting September 1 and continuing through until December. In January the foals are freeze-branded—a definite improvement on the old hot-iron method—with the distinctive MRH (Michael Robert Hamilton) brand.

Some five or six months after birth, weaning takes place. Early on, we established the practice of moving mother and offspring from their more isolated situation into larger groups. This allows the foals to socialise with their peers and aunts in a natural unhurried manner, far less traumatic for them when they are finally separated from their mothers, who are gradually removed, two or three at a time. Once weaned, the yearlings are stabled, handled, groomed and exercised each day leading up to the Yearling Sales which take place in Perth in mid-March. Then there's the breaking and training for the track followed by the excitement of the wins and disappointments at the losses. And so the year goes around: a perpetual string of happenings.

So far as the daily chores are concerned—although everything gradually became more streamlined—a day on the stud today is not much different from what it was nearly thirty years ago. Except of course there weren't the numbers of horse and office staff, gardeners, farm hands and maintenance men back then! Most of it fell to me and a couple of lads. We had a mini-moke with a trailer which the lads would load with feed. I drove and they would feed and check every horse. Later we took the hay out. All this was repeated for the evening meal. And of course there was always plenty to do in the way of picking up manure, cleaning out water troughs, and picking up sticks and stones in the paddocks. The foaling season made it all busier than ever. But I loved it and that never changed.

Over the years there were more horses, people, buildings…and more horses meant increased vigilance, more visits from the farrier, dentist, and vet. As numbers increased there was always some treatment going on—for sore eyes or cuts or colic—especially during the stud season, and all part of the business of owning a stud.

But apart from the welcome modernisations and modifications which brought us such luxuries as the floodlights and twenty-four hour watch in the foaling-down yards, as well as such aids as automatic water troughs and a 'walker'—a sort of merry-go-round which can take up to ten horses at a time on a walk, trot and turning exercise routine—the chores are much the same.

Twice-daily the horses are fed and checked; each morning the manure is collected. In summer the dung beetles are a great help in this regard. Compared to the African variety, these are tiny, and it's quite a sight to see the little insects rolling lumps of manure many times their size—and a great relief to observe the speed with which they dispose of piles of dung. Interestingly, it took one entomologist six years to persuade the Australian government to allow the controlled importation of dung beetle eggs from a laboratory in Pretoria, South Africa, and since then we have never looked back. In the early days, the smaller bush flies rather than stable flies were the main fly problem at Heytesbury. We would hang traps baited with the most awful stinking fish to combat this. Now the dung beetles do a lot of the work and the flies have not been as prolific since. But picking up manure is still a daily task and we have always been meticulous in this regard. Any manure the beetles can't cope with is transported to a 'midden' well away from the yards and paddocks where it eventually breaks down for use on the gardens. There is little waste here: everything possible is recycled and utilised in some way or another. From manure collecting to smoko, the day's work follows a pattern.

It is the more capricious strokes of fate that tend to throw everything out of order and leave confusion or a sense of disturbance in its place. Back in the early Seventies, we certainly had our fair share of catastrophes, some causing a great deal more trauma than others.

As Robert had envisaged, Heytesbury had quickly proved to be an ideal environment for stud purposes and the bush telegraph worked overtime as visiting mares were booked in to foal. Meanwhile he was always on the lookout for opportunities to make additions to our own stock. Our early purchases included two colts—Monoptic and Fair Billum—which we bought at the Perth Yearling Sales. Eventually, Fair Billum opened his winning account at Pinjarra Races and it was not long before they both went on to win in Perth, wearing the Heytesbury racing colours, maroon with a white Maltese cross.

Perhaps one of the saddest events of those early years was the death of Monoptic who died long before he realised his potential. He had an attack of colic the night after he had been gelded at a veterinary centre. After the procedure, it had become obvious that he was not well and Robert had asked the vets if I could stay with the horse overnight. They refused outright, saying adamantly, 'If we do that we will be creating a precedent and then every owner will want to come and stay overnight'. As it turned out, they locked up the clinic at about 6.30 p.m., left him alone all night and by the time they came in the following day he was dead.

By the time I arrived at the centre that morning—Robert having asked me to visit the veterinarian responsible to get a report on exactly what had happened—to my horror they were already making arrangements to send the horse to a knacker's yard. I rang Robert immediately and he insisted that I arrange to have Monoptic's remains brought back to the stud for burial. Never before had I seen my son so upset. When we arrived back at the stud to decide on a suitable burial site, he would say nothing and his face was tense with grief.

There was nothing in the wind to let us know that, all too quickly, we would have to face another calamity.

In fact when Robert came back from the Sydney Yearling Sales in 1974 with the news that he had purchased a two-year-old colt that he had decided to name Haulpak in line with his current takeover interests, and that he planned to race the black stallion before retiring him to stud, we had no inkling that this second métier was fated to take place sooner rather than later.

As usual Robert's judgement of a good horse could not be faulted. In the few times Haulpak raced, he performed brilliantly. Of his seven starts at Ascot, he achieved four wins and three places, returning stakes of over $13,000 on the $17,000 that Robert had paid for him. He had that special ingredient—that elusive will to win—without which all the breeding strategies in the world will not make a successful racehorse. All he wanted to do, and later all his progeny wanted to do, was to lead the field. As it turned out, although it was a tragedy that his first career was to cease so abruptly, he went on to make a resounding success of his second. He was not only destined to stand beside Silver Knight for fourteen years, but also to make headlines for the stud.

The accident happened without warning. Horses leave for their daily workout on the racecourse early in the mornings as there is a requirement that they are off the track by 8 a.m. Frequently, they make their way there while it is still dark. This particular morning, Haulpak took fright at the high-beam headlights of a truck travelling at speed towards him. He shied, slipped, and fell heavily on the tarmac. When the immediate shock had passed and the stallion had been calmed it was obvious that the main damage was to his near foreleg which was seriously injured.

Although there was never any question of putting him down when, after many months of treatment, he had recovered to the point where he was discharged from the vet's stable complex, what a sorry sight he was. His black coat had turned a washed-out bay colour, every rib on his body stood out, and his leg had swollen to enormous proportions. For the rest of his life he would have to contend with a lame foreleg.

Obviously Haulpak's racing days were over. But although he was lame, he was not in constant pain and when his leg seemed to worry him, it was immediately treated. We decided to see what would happen if we put him to stud. And so, rather tentatively at first as we did not know whether or not he would be fertile after the enormous amounts of medication he had been given, we put a few mares to him and charged a nominal $250 stud fee for the service. This sum was to prove unbelievably 'reasonable' in time to come.

Although all this was part of a learning curve, in those days I sometimes felt it was so steep we were moving backwards rather than forwards. The horses were only part of the equation; we had our share of human upsets, too, with staff, for example. Apart from an excellent vet who visited the stud twice a week, and two lads who rode in each day from a neighbouring farm on their bicycles, the day-to-day work and running of the stud was done by myself alone. It became clear that I needed more help.

I interviewed two young girls from France who were on a 'working holiday'; that is, they were travelling around Australia, picking up jobs here and there to pay their way. Although it was obvious that they would not be around in the long term, they appeared very keen and looked as though they would be promising workers around the horses. On the first day they arrived for work, they were not exactly dressed for stable duties, attired as they were in immaculate jodhpurs and well-polished boots, but the next day, they were more appropriately clad in jeans and put in a hard day's work. I kicked off my boots a little earlier than usual that night.

However, my relief was not destined to last for long. On the following day, the two girls, now city-dressed, appeared at my cottage door saying, 'We not here to pick up shit—now we leave…' And promptly walked off down the drive. I must admit that it left me feeling somewhat nonplussed. In fact, I still think it a little odd. How many people on working holidays come equipped with jodhpurs and boots and what on earth had they expected from working in a stable environment? Had they anticipated spending their time cantering over the countryside enjoying themselves?

Not long after this a young lad came to work for us during the school holidays. For most young people, getting away from the city smoke and working in a stud or farm environment is bliss and for this lad it was no exception. All went well until, late one evening, he decided to refuel the tractor. By this time, the regular staff had gone home and my ridgeback Rachael and I were doing a last check of the horses for the night. It was a time of the day I had grown to love at Heytesbury. The late sun shafted through the rapidly growing gum trees turning the backs of the grazing horses to gold and there was a sense of peace and oneness with the world that was not there during the busy days.

I was offering a carrot treat to one of the horses when I was startled by the most hideous scream. For a split-second, I stood still, stunned. Rachael yelped. Language followed with which, I am glad to say, I was not familiar. I raced in the direction of the commotion, which turned out to be the tractor shed, and to my utter horror saw the lad in flames. He was wearing only a pair of rugby shorts which were on fire around his body. The tractor, too, was burning like a furnace.

Fortunately a neighbour was passing our entrance and hearing the commotion had rushed in to assist. So while I raced to my car to grab a sheet to wind around the poor lad, the neighbour ran to a pile of sand that had luckily been delivered that very day and filled bucket after bucket to throw over the flaming tractor. Meanwhile, I half-carried, half-pulled the boy into the car, switched on my headlights and broke all speed limits to get him to the nearest hospital in Pinjarra.

The full story was short and contained a lesson that, sadly, the lad would never forget. Apparently, he had omitted to turn off the ignition while he refueled the tractor which almost immediately burst into flames. It was a painful and terrifying experience which resulted in third-degree burns over 90 per cent of his body. Every time I see the burnt rafters in the shed it reminds me of that ghastly day. In those days, both the diesel for the tractor and the fuel for the cars were contained in drums. The accident led to the installation of petrol and diesel pumps, although this of course would not have prevented its happening.

Not long after this terrible incident, I had another, quite different experience with a fellow whose parents wanted him to work at the stud to enable him to decide whether or not he really wanted to work with horses. It is the nature of the horse industry that it attracts young people who think they want to work with horses until they find there are many menial jobs to be done, and often long hours to put in when there is a sick or orphaned foal needing constant attention. Many also find that they are some distance from the bright lights and there isn't much in the way of entertainment available after working hours. The reality of the work is often at odds with the dream.

Anyway, this young chap seemed very keen and we were prepared to employ him for a month on a trial basis. But just before the end of this period, he failed to turn up for work one morning. After waiting around for a while, I went up to his quarters but there was no sign of him. At noon, he was still missing. I rang Robert who suggested I call the police. We were worried that he had been involved in an accident. His parents lived in Singapore and there were no relatives in Perth whom we were able to contact. Later that afternoon, the police arrived and searched the property for clues. It was nerve-racking to watch them walking around the dam and peering into its muddy depths. In fact, the uncertainty alone was distressing to say the least. The police eventually left, but we were still no closer to finding a reason for his disappearance.

Later that night, I was startled by the sudden ringing of the telephone. I suppose I was already on edge after the worries of the day. It turned out to be the local hospital calling to let me know that the lad had been admitted the previous night. Apparently he had had some form of poisoning which had rendered him semi-conscious and he had been picked up from the local pub by a man who, not knowing his address, had done the right thing by taking him to hospital. It was some time after he had his stomach pumped and medication administered that he was able to communicate. The next morning, I fetched him from the hospital and dragged the story out of him. It seemed that at home in Singapore he was quite used to eating a particular variety of mushroom. Recent rains had encouraged a flush of fungi around Heytesbury and these he had gleefully collected and eaten. The result was food poisoning.

Not only had the whole episode been worrying, but it was unnecessary and time consuming, and even more annoying when the lad did not even apologise for the trouble and upset he had caused. Needless to say, that was the end of the trial period.

These incidents aside, right from the beginning Heytesbury proved a marvellous place in which to bring up children. It was, in short, a child's paradise. The families at the stud were able to take advantage of the same opportunities my grandchildren had. There was no need to worry about the dangers of roads or fast cars, and they learned to ride bicycles and horses, drive a car and a tractor, and swim or fish for marron in the dams. And of course there were the animals, both their own and those of the stud. It was like having a farmyard to play in: there were plenty of soft, furry animals about, both large and small. It was not long before we decided to keep at least two horses for the young people to use for hacking. And as the staff grew, inevitably so did their pets—quite a mixture of dogs, cats, birds and rabbits.

One family had a sheep that followed them about like a dog, which reminds me of an evening when my ridgeback and I were out for a walk. It was one of those occasions when I would have given anything to have a camera with me. Christine, who eventually worked as the office administrator, her husband Garry and their family were out walking with their extended family. This was headed up by a docile, rather well-fed sheep which was followed by a proud-looking cat which stalked with its tail straight up in the air. Two small dogs trotting along at the rear and the small tame bird on Garry's shoulder completed the picture of a somewhat unusual but particularly happy family.

This was the background against which Robert and Janet's children grew up. They regarded the daily business of the stud as a normal and natural part of life and it proved to be an ideal environment for their later careers.

The two elder children, Peter and Catherine, had been born by the time the stud was purchased, Peter in 1968 and Catherine in 1969. Simon followed in 1972 and Paul a year after that. Almost every weekend Robert and Janet would drive down to the property and, just as I had in Africa, all four children learned to ride, swim and drive a car on the stud. What bliss the original four hundred acres of undulating farmland and unlimited opportunities for horse riding and camping would prove to be for four energetic siblings.

Meanwhile Heytesbury continued to grow. Robert worked fast. He started acquiring extra land. In 1975, the adjoining farm came up for sale and was purchased. In this acquisition, there was, however, a snag. It was a dairy farm and part of the agreement was that we were to run the dairy for two years to meet the Dairy Board's milk quota agreement. It proved to be an enormous amount of work and something that in hindsight all of us would have preferred not to have taken on. The cows had to be brought in very early every morning for milking, seven days a week, winter and summer. And this all had to be finalised by the time the milk truck arrived to load up the old metal milk cans at 7 a.m. This sort of schedule combined with the night work of the horse stud, particularly at foaling time, was, quite honestly, punishing.

It was at this point that the need for more staff again became imperative. Robert employed a man who took over some of the maintenance work and the more general running of the property. Another man was made responsible for the irrigation system, and soon the load was a little more evenly spread. We also took on several young girls to help with the ongoing studwork. I was beginning to find out what running a stud was all about.

Myself, Cape Town, 1944.
Taken around the time Peter was posted from the Mediterranean to South Africa.

Ronnie (right) at Sandhurst in 1937.

Polo team in Northern Rhodesia after the war. The fourth member of the team can be detected at Ronnie's feet.

Ronnie with Frankie, the leopard cub reared by Erica.

Robert riding Family of Man on one of his weekend visits to the stud in the early 1980s.

Heytesbury yearlings enjoying their freedom.

A visit to the stud. (Left to right): Paul, Catherine and Simon, with Paul's golden Labrador, Charlie.

Heytesbury as it is today. Taken from the top of the Darlings looking toward the coast.

Initially I had used a stable as an office. It was pretty basic, but comfortable, and contained all the essentials like a filing cabinet, a deck chair, a small table and the portable typewriter which had travelled the world with me. We even had a telephone, although at this point it was only a party line—which meant we had to dial the telephone exchange for a connection—which closed down at 9 p.m.. We used a card system for our records, which I nostalgically hold on to, outdated as it is in this electronic era.

But as the stud developed, it was decided I was to have a real office. An office block was built with three offices, an area for my secretary/receptionist, a large kitchen, washrooms and an area for entertaining. And luxury of luxuries, the party-line was replaced by a 'modern' telephone which allowed us to simply pick up the receiver and dial directly instead of going through the exchange. By this time, too, we were connected to the scheme water supply, which was useful for emergencies, although by now we had the dams. Next the electricity was laid underground. Heytesbury was growing.

Robert didn't stop there. It was not long before additional land was purchased. He acquired the farm across the highway, which was named Little Heytesbury and was to be used mainly for the rearing of yearlings. More acreage was bought which stretched further along the highway. This was when we decided to make a new entrance for the stud at what was known as the Forty Mile Peg leading off the Southwest Highway.

One change led to another. It now became obvious that the best place for the stud office would be at the new front entrance which happened to run alongside the original foreman's house. And so the house was converted to an office and this is where the stud office continues to stand, surrounded by lawns and native flowering shrubs in the natural Heytesbury fashion. Electronic wrought-iron gates set between tall limestone pillars monitor the comings and goings of the stud—a far cry from the original entrance to the property which was off Boyd Road and marked with a timber sign with the words Heytesbury Stud carved out of the wood and painted in white.

The gradual increase in staff necessitated additional accommodation, so plans were made for building a block of single quarters for the single staff and detached cottages for those who had their families with them.

While all this was going on for the benefit of staff and the stud in general, the horses certainly weren't neglected. The first stable complex had been built and the property had been divided into paddocks and foaling-down yards. We built five large yards for the waiting mares, and six individual foaling-down yards directly in front of my cottage. I could now keep an eye on the initial proceedings from my sitting room or study windows—a definite improvement on waiting in the car on a cold winter's night as I had for the birth of Tixall's colt. The floodlights that had replaced torches were certainly a great deal more efficient at monitoring a mare in labour. Each evening at sunset during the foaling-down season these are switched on and the whole area looks as though it is lit by a huge full moon.

One of the more significant additions was the sick bay block in which we had installed infra-red lights and walls lined with timber. The lining was originally added to this building because we had a mare with a broken shoulder who would use the wall as a lever to get to her feet.

Although the barn has been used many times for operations, a particularly difficult case was a caesarian we were forced to perform on a rather well-bred mare in the middle of one of our early foaling-down seasons. It was about 4 o'clock in the morning when the night-duty person sounded the alarm and in a matter of minutes we were all rather anxiously assembled: the vet, the night-duty staff and myself. This was a visiting mare in foal to Haulpak, so both mare and foal were very valuable. Part of the problem was that the foal was badly positioned, presenting its hindquarters instead of its head for birth. This does sometimes happen, and we usually deal with it quite successfully by pushing the foal back inside the mare's body and turning it around. But in this case, the problem was compounded in that it was a particularly large foal (as almost all Haulpak's offspring tended to be). In a very short time it was obvious the mare was in trouble and we ended up having to give her oxygen while we literally cut the foal out.

Dawn had broken and the sun had started its rise above the ranges by the time we finished. As it turned out, we lost the foal but saved the mare. These were the years in which we did a lot of learning.

CHAPTER SEVEN

The 'Old Poacher'

It was April 1978 and I had been widowed for over fifteen years. It was a beautiful autumn morning, sunny, but without the big heat of the West Australian summer—the sort of day to think and potter. This particular day I was in the process of creating a waterfall and pond in my garden.

Quite apart from the progress of the stud, I was feeling very proud of what I had achieved over the six short years I had been in the cottage. The lush lawns were well established and provided a vista of restful green for as far as the eye could see. Placed randomly across these lawns, stately eucalypts reached up into that incredible blue of the Australian sky. The eucalypts had originally been planted to shade the first foaling-down yard and to shelter the cottage, but now they were so huge they were almost overpowering and looked as though they'd been there for a hundred years. I have seen many many paintings of gum trees, but never has the artist been able to capture anywhere near the intensity of colour found on the trunks of these trees: bark of the most magnificent shades of green, rust and grey that become even more vibrant when it rains.

The waterfall was being constructed just to one side of this scene, far enough away from the cottage to draw the eye, close enough to hear the splash of the water while sitting under the pergola. I was building it with the help of an excellent contractor who had collected some attractively marked rocks from the top of the ranges. The earth had been excavated, the depression cemented to make a pond and to one side of this the rocks were piled one on top of the other to create what was above to become a waterfall.

My younger son Simon had mysteriously disappeared in the Tsitsikamma National Forest, in the Cape Province in South Africa the year before, and I had in mind to make this into a memorial for him. He had been a peaceful person passionate about nature and I felt that this water garden represented these characteristics. As we worked away in the quiet warmth, I could almost hear him saying, 'That's really great, Mum, thanks'.

My thoughts were far away when Joy, a friend from a neighbouring farm, arrived with the day's *West Australian* and my mail. We stood in the garden, enjoying the sunshine and watching the workers as the waterfall took shape. It was not until I went to make a cup of tea that I glanced down at the mail in my hands. On top of the rest was a pale blue airmail envelope addressed in Ronnie Critchley's strong handwriting. My expression must have changed because Joy, glancing first at my face and then at the letter, could not refrain from saying with a frightfully cheeky grin, 'Another letter from Forfar? Dare I ask who is the Scottish admirer?'

Women will always enjoy love stories, their own and other people's, and anyway I was glad of a chance to talk about the past. So I said, 'Well, it's quite a lengthy story of long ago, but if you'd like to hear it, I might as well put you in the picture!'

We decided to sit on the verandah under the pergola with a pot of tea. Now the contractors had finished their work, I was able to turn on the pump and set the little waterfall running over the rocks. It was every bit as soothing as I had anticipated and we sat there in the warmth very happily.

'Where to begin? I asked.

'At the very beginning, of course. And photos. Show me lots of photos,' said Joy.

She certainly asked for it! I collected my albums and a framed photograph of my first husband Peter Holmes à Court, because the story I had to tell began just a little earlier than when I had first met Ronnie. It began in the war years when the children and I were living in Southern Rhodesia while Peter was serving in the Royal Navy and away at sea. Peter's ship, the *HMS Delhi*, had been badly bombed towards the end of the war and he was severely shell-shocked.

The framed photograph that I handed to Joy of my handsome husband in his naval uniform was taken just before the war ended. I also had a rather yellowed newspaper clipping of the *HMS Delhi* bombing in the Mediterranean, which reported the unhappy details of one of the war's awful disasters.

'Peter was lucky to survive,' I said to Joy, 'but with hindsight I suppose it was almost inevitable that he would carry the shock of that war for the rest of his days. His health was never the same. He was in an extremely fragile mental state and suffered from blinding headaches and frequent nightmares. Sudden unexpected noises would set off panic attacks.

'Consequently, he was posted to Simonstown, near Cape Town, where it was hoped that the temperate climate would help his recovery. I hoped that the joy I felt being with him and the love that came from being a family again would complete the recovery. It certainly helped. Although Peter would never be the same, those fifteen months we spent in the Cape were the happiest we spent together.

'When he was demobbed in Salisbury at the end of the war, we were lent a flat in Rotten Row near the centre of the city and we tried to settle down to family life. It was at around this time that I first met Erica, a woman who was to become one of my best friends and to whom my life would always be linked in rather an unusual way.

'It was 1947, the Royal Family was due to visit and our flat had a rather handy balcony overlooking the main street down which King George VI, Queen Mary and the two princesses, Elizabeth and Margaret, would be driven.

'Because Erica had horses racing at a Salisbury meeting that weekend, she was down in Southern Rhodesia from Chikupi— her ranch outside Lusaka in Northern Rhodesia—and decided she would like to view the royal parade. She asked her trainer Len to find her a good spot.

'I knew Len through a mutual friend Beryl, with whom I eventually ran my riding school, and he had no hesitation in asking me if they could make use of our balcony. That was my first meeting with Erica, a vivacious whirlwind of a woman, not much taller than myself, but whose quite outrageous style topped anything I had seen before…or have come across since, for that matter. The language she used would have stopped a sailor in mid-stride!

'Our friendship began the moment we were introduced by Len saying, 'I'd like you to meet my client Lady Fitzgerald…' And, although I didn't know it then, it was a bond that was to continue for twenty years. A glimpse of Erica's determination as a young girl probably gives a greater insight into her character than anything else. The story goes—and Erica certainly gave the country an excuse for plenty of stories—that as a young woman she went to a ball and across the room she saw a tall, rather charming-looking man, a Sir William Fitzgerald, who was one of the country's leading legal experts. On a visit to the cloakroom, Erica stated aloud, 'I'm going to marry that man.' And marry him she did, although the marriage came to an end when Sir William went to Palestine and, after a brief visit, Erica decided she did not like it and returned home.

'Erica's second husband was a rather long-suffering fellow called Michael Lafone. That marriage wasn't destined to last much longer than the first and it is testament to the amount of stress it produced that, when it finally broke up, Michael moved from Chikupi to a cottage in Lusaka, which he rather aptly named Wit's End.

'We were of course to learn all this later. All I knew at the time was that we had formed an instant liking for each other and exchanged a great deal of information in a very short time. Below us, the royal parade came and went. The cheering died down. Erica had been standing on our balcony taking rather generous sips of champagne. She looked at her watch, 'I must go.' She threw her head back and drained the glass. 'But don't let's lose touch. Why don't you come up for the Lusaka Show one day. And of course you must stay with us at Chikupi!' With a cheery wave of her hand she was off, Len close behind her.'

Joy was still listening, asking questions from time to time as I passed her a photo, and our cuppa turned into a sandwich lunch as I continued the tale.

'And this is where Ronnie comes into the story. Ronnie Critchley was a retired cavalry officer who was attempting to put as much distance as possible between himself and a traumatic second marriage. Makeni, a cattle ranch of 19,000 acres ten miles from Lusaka, was a neighbouring property to Chikupi, and not only gave him the space he required but also the opportunity to try something new, a challenge, which was something Ronnie always found appealing.

'His move to Makeni followed his retirement from a very distinguished career in the army. He retired with the rank of Lieutenant Colonel and was much decorated. He had been awarded the DSO (Distinguished Service Order) in Burma, the Military Cross and Distinguished Military Medal Emperor Haile Selassie for services in Ethiopia, and the Insignia of

Honour in Zambia.

'In Rhodesia the vast distances between properties meant that neighbours invariably became very good friends, and entertaining those friends became a way to beat the solitude. Makeni and Chikupi were neighbouring properties so it was not long before Ronnie became a regular visitor at Erica and Michael's house.

'Erica simply adored entertaining. She was a bundle of vivacity and loved people, men in particular. Not only did she provide the neighbouring towns with plenty of gossip, but she was also a great raconteur. Her sundowner parties would begin at seven in the evening and invariably consisted of several rounds of drinks before dinner was served at around 10 p.m. However, poor Michael, usually fairly intoxicated by then and having heard Erica's stories a number of times, quite often fell asleep, sometimes with his head on his plate.

'Because of their mutual interest in wildlife (although to be honest, at this point Ronnie had done more killing than preserving to the extent that he sometimes referred to himself as an 'old poacher') Ronnie and Erica very quickly formed an association and before long they started the Wild Life Society. The society had branches in all the main towns in Northern Rhodesia and it was not long before there were over 1,000 members. They started a magazine called *Black Lechwe*—a type of antelope—which very soon had a flourishing circulation.

'By now Robert and Simon were down in South Africa at boarding school and I had started my riding school. When the boys came up to Rhodesia for the school holidays, they not only did their bit to help out, but were becoming increasingly skilled and experienced riders. Robert was a particularly confident rider and soon we were not only competing at horse shows around the country, but winning just about everything we entered.

'It was July 1949, with the Lusaka Show coming up. We decided we would compete that year and that we would take Erica up on her offer of accommodation at Chikupi.

'My first sight of Chikupi? It was early afternoon when we drove up a winding driveway between the tall straight mopane trees and wild figs which the animals tended to enjoy with somewhat drunken results. We drove right up to the huge rambling homestead and stopped in front of the long verandah. Everything about Chikupi was large, from the enormous logs stacked neatly on either side of the open fireplace to the huge rooms, the grand dining table, the guest cottages…and Erica had all the servants in the world to help her with it. She ran a fine house, and her parties were sensational and frequent, held week after week until the early hours when guests would switch to drinking whisky and milk.

'I emerged from the enthusiasm of Erica's greeting to see a tall slim man dressed in an open-necked checked shirt sitting back in a chair apparently contemplating the gardens from the shade of the verandah. Ronnie Critchley rose to introduce himself just as Erica stopped her flow of talk long enough to ask, "Do you want gin or tea?"

'At the Agricultural Show, we won everything we entered. Peter was riding a horse called Gentleman, Simon was on Honey, Robert on his Mr Jeremy Stickles and myself on Jonathon. Erica lent Robert a pony called Toshey for the Best Child's Pony event and was so impressed by the way he rode she ended up giving it to him.

'We went back to the ranch to celebrate our wins and the evening rocked on until Erica suddenly remembered that she was supposed to present the prizes and cups at the show ball. So in a less-than-sober state we hurried to the ball arriving at 11p.m.…in the nick of time as it turned out.

'Time rolled on. Erica and Michael finally divorced and then, perhaps inevitably, because of the intensity of their mutual passion for wildlife, Erica and Ronnie were married some eight years after they had first met. Or as Ronnie put it, "I drifted into marriage with Erica". Although Chikupi had been Erica's family home, she sold it at this point and went to live with Ronnie at Makeni until, in turn, that was sold and together they bought Blue Lagoon.

'By this time, my marriage to Peter had sadly come to an end. I went through an extraordinarily emotional few years until I met a civil engineer called Charles Trevor. It was not long before Charles became increasingly involved in my riding school and we formed a business relationship with the aim of making Rydal Court, as it was called, the foremost riding school in Rhodesia. We were a great team both on and off horseback and one thing led to another until we decided to marry in 1956, the same year as Erica and Ronnie wed.

'By now Erica and Ronnie had set up the Wild Life Society in Lusaka. They were to run it for the next twenty years; he as president and she as secretary. Because I was every bit as passionate about the preservation of endangered wildlife and the rearing of orphaned animals in that part of the African bush, for the rest of Erica's life this mutual concern infused the letters that we continued to write to each other. As it turned out, she was to remain at Blue Lagoon while my travels took me all over the world.

'When did I last see her? It was in 1962 when Charles and I were in the middle of shaping our Chobe dream—the hotel we built on the banks of the Chobe River in Bechuanaland. At the time we had a pesky monkey called Jeremy—an orphaned vervet we reared after he lost his mother—who started off his life being cute and rapidly became a pest by literally terrorising the hotel residents. He would scamper over newly-set white tablecloths with his dirty paws, steal paperbacks right out of the hands of the guests and help himself to money from the till. Our efforts to release Jeremy into the bush were useless. He was having too much fun at Chobe and kept coming back for more.

'In the end I sent a somewhat exasperated *cri de coeur* to Erica and she suggested we bring him up to the ranch. It was some time since we had thought of having a holiday and quite apart from our problems with Jeremy it would be good to see Erica and Ronnie again. So we popped Jeremy in a cage, much to his undisguised and extremely voluble disgust, and took off for the north.

'We drew up in front of the homestead in the late afternoon when the sun was tilting through the trees. At first I thought nothing had changed…and perhaps it hadn't really. There was Ronnie walking across the lawn to the swing seat where he sat rocking gently to and fro in his usual quiet way. A servant, impeccably dressed in a white kanzu and red fez, brought out tea and a sponge cake which he set down carefully on a table in front of Ronnie. Perhaps the only difference was in Erica; although she held the floor in her usual way, she seemed slightly less outrageous, just a little more circumspect. Ronnie has tamed that woman, I remember thinking at the time.'

I stopped. So much had happened from that point to this and it was impossible to cover it all. Joy's attention hadn't wavered since I started. She prompted me. 'I know about Chobe and your coming here. But what about Erica?'

'Very sadly, she died of cancer a year or more ago, around about the time Simon disappeared. Ronnie wrote to let me know, and our correspondence since then has helped both of us to deal with our losses. He's living in Scotland at the moment. And from his accounts of his hunting and fishing expeditions, it sounds as if he has reverted to type—the 'old poacher' has re-emerged. Understandably, Blue Lagoon was far too depressing and lonely without Erica and he needed the complete change.

In his last letter, he mentioned he was thinking of coming out here for a holiday at some stage.

'And so perhaps you *will* get to meet him! But until I read this,' I waved the letter aloft, 'you know as much as I do!'

CHAPTER EIGHT

But Can He Cook?

By the time Joy leaves, the day has all but gone. The rays of late afternoon sun send shafts of light shimmering among the top branches of the tall gums and for an instant my mind harks back to the sun through the leaves of the mopane trees at Makeni. Here the softening light is my signal to start thinking about my evening tasks and the carrot treats for the horses.

But I still have to plant out the pond. The bucket of muddy water lilies is lying where I left it outside the kitchen door. Before the horses, before reading Ronnie's letter, I decide to plant just one, a yellow lily. Bending at the edge of the pond, gently untangling the roots to allow them to sink into the soft mud, spreading the leaves to float on the sun-spangled surface, I place the plant in its new home. Sit back on my heels to admire the result. Water games. My boys. Yellow is the colour of the mind. Tracking back over time has unsettled me. I am still in the mood for reminiscing.

Emotionally it had been a turbulent year and I was still trying to find my balance. First the news that Simon was missing in May, 1977 was followed a month later by the news that his Datsun pick-up van had been found abandoned deep in the Tsitsikamma Forest that runs along the coast of the Cape Province. By September I could no longer bear the uncertainty and unsatisfactory nature of the investigation and flew to Southern Africa to speak to his friends and help the police in any way I could. Although a few pieces of a skeleton which matched the dimensions of Simon's frame was found three years later, over the next quarter of a century all sorts of conflicting information was to emerge. Everyone who became involved with the quest to find Simon, and there were many, had a different story.

I adored both my sons. They were so completely different from each other that there was never any temptation to compare them. Robert was the one who sported a clean white handkerchief in his pocket; Simon, a carpet snake. Each had unique qualities. Simon was a particularly private person, gentle and kind, with an extraordinary empathy for animals. Paradoxically, he was also an adventurer in the best sense of the word. His job as senior game warden in the Bechuanaland Protectorate had allowed him to exploit both sides of his nature to the full. But his death made no sense. It was inconceivable that he would have committed suicide, that just wasn't Simon. At the same time, no motive for murder had been uncovered. An accidental death was possible, but why then would the licence plate have been removed from his abandoned van; why would the engine number have been scratched off? Part of me had never been convinced that he was dead. There was no real evidence and he always was inclined to disappear. With intent to photograph one animal or another he often disappeared into impenetrable *bundu* for months at a time. But the pragmatic side of my nature reminded me that months were somewhat different from years. So I continued to vacillate back and forth, one minute thinking he was still alive, the next uncertain. The death of someone is difficult enough to cope with, but losing someone and not knowing whether they are alive or dead is ten times worse.

Ronnie's first letter telling me of Erica's death reached me in the middle of this turbulence adding to the distress of that year. Why is it that one sadness seems to attract another? My wonderful friend had died of inoperable cancer after twenty years of tireless work on behalf of her beloved animals of the wild. Following her death, Ronnie returned to Blue Lagoon, the 100,000-acre wildlife sanctuary he and Erica had established at the centre of the Zambian hinterland.

It was hard to imagine someone with so much vivacity, so much verve, no longer being alive. It is certainly a comforting thought that this energy forms the spirit that, it is said, never dies. When I think back to the last time I saw Erica, I don't only see her small energetic figure waving goodbye from the verandah of Makeni Ranch. I also remember her complete fearlessness of wild animals. Once I watched her get into a cage with Frankie, a young orphaned leopard—whose life she saved before releasing him back into the wild—only to come out minutes later with blood oozing from deep claw marks down the length of her arms. It was the cub's idea of play and her thin busy sun-browned arms were covered with graphic scars.

Ronnie's letter went on to say that she had been given an appropriate tribute for her work. Her ashes had been scattered from a Zambian Air Force helicopter over her old farm Chikupi. The Blue Lagoon National Park had been purchased by the Zambesi Government as a national monument to be maintained in her honour. Sadly, although the house was left as a shrine and was to be kept in immaculate order, I suppose it was inevitable that under then President Kaunda it would deteriorate. When I visited some years later, army bunk beds had replaced the grandeur of the old homestead, the wildlife was once again being killed and the beautiful mopanes chopped down for firewood.

Erica had been a close friend, the first person I had phoned on the death of my husband Charles, and her reply at that time had been immediate and unreservedly compassionate. She had invited me to Blue Lagoon to recover from the shock and loss, but the sale and winding up of the Chobe River Hotel and the necessity of dealing with the trivia of day-to-day life had intervened. I wished now that I had gone. Or were the memories I had the right way to remember her? One thing was for certain, the loss of her vivacious personality would leave another gap in my life. In a different way, Ronnie's loss was also mine. I wrote back in empathy with his barely concealed pain, at the same time telling him of Simon's disappearance.

Some time elapsed before his next letter. Was I surprised when it came from Scotland? I don't think I was.

It had proved much too difficult for him to continue living at Blue Lagoon, the ranch where he and Erica had lived so happily for so many years. Day and night the memories had pressed in on him until the loneliness finally became unbearable. He returned to Scotland to find somewhere to live and allow his emotions to settle. Not only had he lost Erica, but over thirty years in the heart of Africa. He needed to spend time somewhere entirely different, doing entirely different things. Perhaps that's why the wildlife conservationist of the past twenty years regressed.

To understand these two urges—the hunting and the conserving—both of which he did with enormous skill and unremitting voracity, is to understand Ronnie. I remember his admission, 'When I first arrived in Northern Rhodesia in 1948, I really was an "old poacher". I had spent most of my leisure time in my cavalry days shooting duck in the Middle East and chasing ibex and stag in the Himalayas. Now it was time for me to put something back, to give up the gun, and try to look after the wild things which had given me so much fun in the past.'

It was typical of Ronnie that his idea of 'putting something back' didn't stop at pursuing the African poachers with every bit as much tenacity as he himself had once applied to hunting. When he and Erica married, he sold Makeni with its 3,000 cattle and they moved to Blue Lagoon—over 100 square miles of wilderness on the edge of the Kafua Flats—seventy miles outside the Zambian capital of Lusaka.

For the next twenty years they lived in a literal wilderness with the nearest European neighbour 40 miles distant and the local African population outside the ranch boundaries numbering only three to the square mile. Twenty years spent among wild animals and hundreds of species of birds. All that time spent together saving the orphans of the wild—elephants, lions, leopards, cheetah, hippo, bush pig, and many species of antelope—from death. At the end of each rainy season, the River Kafue would burst its banks, flooding a third of the property, the waters so high that it drove the red lechwe to the front door of the ranch house. Waking in the noisy stillness that comes with an African dawn to find fragile lechwe fawns seeking shelter on the red concrete of the verandah. Why did it come as no surprise to me that the memories were too strong for him to stay?

'Operation Noah'—a wildlife rescue operation—was perhaps the greatest project that Ronnie and Erica undertook. At the time, in 1959, the plight of the animals stranded by the damming of the Zambesi River made headlines in newspapers around the world. The problem arose as a result of the building of the great Kariba dam, an enormous water catchment that straddled the boundaries of Northern and Southern Rhodesia. In the building of the dam, neither government had given a thought to the plight of the wild animals. The Zambesi River, for 50 miles from the dam to its confluence with the Kafue River, had been reduced to a fraction of its normal dry season flow while the waters upstream on the other side of the dam rose correspondingly. Animals marooned on the fast disappearing islands began to starve and drown. Thousands of tiger fish were found dead with their stomachs full of crickets that had been driven from their homes in the sides of the banks as the waters rose.

On the other side of the border, Southern Rhodesia was deeply involved, but a tremendous amount of time was wasted in trying to get the government of Northern Rhodesia to do something. It wasn't until the Fauna Preservation Society in London was alerted that anything of value was achieved and in no time it acted to raise the money and volunteers needed. Ronnie and Erica were in the front line of the rescue operation. They not only involved themselves in the essential fundraising efforts, but sailed to the sinking islands, netting and tagging the animals, loading them onto the boats and taking them to the mainland. There is always a certain amount of risk involved when dealing with animals, and when those animals are not only wild, but hungry and terrified into the bargain, the risk increases one-hundred-fold. Probably nothing says more about Ronnie and Erica's involvement as the naming of the forty-five-foot parent ship built especially for the project. Large enough not only to carry people and equipment but also to tow smaller boats to the scene of the rescue, it was called Erica. I think of it as the first boat specifically built for animal rescue work since the ark.

And this certainly wasn't a one-off operation. Their efforts continued until 1963 by which time all types of wildlife had been saved—among them literally thousands of antelope, zebra, warthogs, snakes—even porcupines which I always associate with England rather than Africa.

But it was indeed Africa, an Africa Ronnie had to get away from after Erica's death. It is natural that he should have returned to Scotland, the country of his birth and about as different from the heart of Africa as you could imagine.

He found a refuge in a little town called Forfar, near Angus, in the foothills of the Sidlaw Hills in the east of Scotland. A rather picturesque farmhouse set in some very pretty country, and obviously chosen for its prolific fishing and shooting possibilities, suited both his mood at that time and his character. He bought himself a pointer from the daughter of the Duchess of Montrose from whom he had purchased his first pointer Jackie—one of the loves of his life—fifty years earlier. He cleaned his favourite gun, set to training his dog, and was ready to fish and shoot—the hills and river valleys of the lush Angus countryside providing ample scope for both.

The conservationist had been placed on hold. The 'old poacher' had reverted to type. Why? Perhaps this was the only way he could put to rest the ghosts of the past. Maybe each time he raised his gun he had in his sights one of the evil spirits responsible for taking Erica from him. Who knows what demons were shattered as the sound of the shot spread through the hills, wrapping itself around the tors of the craggy Sidlaws and coming back to him on the wind? Who knows how much pent-up adrenalin was released each time he raised his gun to follow the canny snipe, ducking and diving above the heather of the moors?

I can only speculate. Hunting was definitely in Ronnie's blood. He grew up with a gun for a toy, shooting rabbits for the pot under the watchful eye of the gamekeeper on the family estate in Dumfriesshire, only seven miles from the border post of Gretna Green. Quite a collection of servants was required to keep up both the inside and outside of an estate of this magnitude; the grand Victorian mansion was tacked on to the side of a sixteenth-century border keep set amid lush parklands that were in turn surrounded by dark-green pine forests. There were no less than fifteen domestic staff including kitchen and pantry maids, nursemaids, a cook, butler and footman, while the outside was taken care of by the gamekeeper, several gardeners, foresters and a dairy maid.

Although each chapter of Ronnie's life was totally different from the next and each was outstanding in one way or another, he would probably describe his beginnings in this easy-living environment as one of the most idyllic. Despite the cold wet climate of southern Scotland, this was where he learned to ride, fish, shoot and hunt. He and the gamekeeper would take off on regular 'ferreting' forays to supply rabbits for the staff, and the sheep in the home park were carefully stalked and shot. Keeping close to his heart the hunter's oath—never shoot more than you can eat—he learned to shoot clean and straight and never to leave a wounded animal in pain. Afterwards or on weekends there would be hunt balls, cricket and tennis parties and constant big dinner parties, all part of the Scottish country social life. Another reason to return to Scotland—to the distractions of his youth.

But now it was 1978, over two years since Erica's death and he had been in Scotland for the past year. His letters had gradually become a great deal happier. For someone of his sporting nature, it was pure bliss to fish for trout and salmon in the sparkling icy streams of Scotland. He shot for grouse over the head of Jackie, his new pointer, in some of the best bird-hunting country in the world. And then it seemed that even this was not enough, he went further afield with Jackie, to the other side of Scotland and over the water to game expeditions on the island of Islay. Then north, as far north as he could reasonably go, both of them ferrying over the icy Atlantic Ocean to the Orkney Islands. He was happy. He wrote. 'The countryside is beautiful and the people most hospitable.' The hunting therapy had worked like a charm.

It therefore came as something of a surprise when he mentioned he was thinking of taking a break from the charms this lifestyle held for him. He was contemplating 'a short holiday in Australia which would', he had said in his last letter, 'give me a chance to renew our friendship'. At the time I had been rather astonished. Knowing only too well how difficult it can be to set aside the pull of Africa and find a compensating lifestyle, I wondered a little at his intention of disturbing the solace he appeared to have found.

But if his last letter had been a surprise, the latest was even more so—and rather sudden I might add. It appeared his intention was to emigrate from Scotland, and not to Africa where he had spent thirty very happy years, but to Australia! And not only to Australia, but to that little haven that was in the process of maturing, Heytesbury Stud itself. If Joy had noticed the change on my face when I spied his handwriting, she should have seen my face when I read that.

Nothing in his letters had given me any inkling that he might do something quite so radical. I admit I felt slightly perturbed. A holiday is one thing, immigration quite another. Why on earth would he want to immigrate to Australia when it seemed he had achieved quite a measure of happiness in his home country? Where would he live? And what would he do here? Western Australian rivers were rather sorry distant cousins to those he had been used to. I put the letter down and thought back to my own difficulties in settling in such a very different country. No, I was more than disconcerted. I was stunned.

I decided to take my time about replying. To be perfectly honest I was in two minds about Ronnie's decision. It was not that I had any say in his coming to Australia, but living with me, as it appeared he had in mind, was a different matter. I was not sure I wanted my life distrupted.

I was sixty-three years of age and had been on my own for fifteen years. I was comfortably living in my pretty gum-shaded cottage and happily involved in the business of the stud. By this time all four grandchildren had grown out of babyhood. They were little people now with definite personalities and I was getting a great deal of enjoyment in encouraging their passion for horses and riding, just as I had with my two sons.

And, significantly, I also happened to have a very satisfactory social life. I was at that delightful point where my adopted country was beginning to become my home and shared experiences allowed acquaintances to start the process of turning into friends. Consequently, rarely a weekend passed without an invitation to a dinner party or a barbecue, most of them in the city, of course. I thought nothing of driving for an hour up to Perth to attend a party and coming home in the early hours of the morning. Or staying over with my hosts for the rest of the night and arriving back at the stud by six a.m. in time to supervise the early morning feeds. Although there is no doubt I had always been attracted to Ronnie, it was an attraction tempered by the long time we had been friends. I was still feeling battered by Simon's disappearance. To tell the truth, I wasn't sure that I wanted my emotions disturbed by a close partnership again.

The only thing that was not totally satisfactory about my lifestyle at that juncture was my everlasting hatred of cooking. Cooking was something I had never mastered and even the idea of making something as apparently simple as an omelette would upset me. Sandwiches I could manage, but I'd as soon saddle an unbroken horse as put a pan on a flame. And then the thought suddenly occurred to me: perhaps Ronnie could cook! I must admit to being slightly ashamed that this seemed to me the only redeeming factor in his coming out.

When the next letter arrived before I had a chance to answer his last, he made it even clearer. Did I want a 'companion'? Cooking aside, that was not so easy to answer. I certainly wasn't sure I did. On the other hand... Obviously, the time had come to supplement our correspondence with a telephone call. I picked up the phone and dialed Forfar.

That was the first of many a long phone conversation and the start of a correspondence that became daily over the next few months. I needed to know so much more about a man who was prepared to gamble the fragile happiness he had so recently achieved to start a life with a new 'companion' in such a completely different country.

CHAPTER NINE

A Shaft of Golden Sunlight

Conjecture became fact. Ronnie did come out. In the end I had about a month's warning. And what a month it was.

For a start, although Robert and Ronnie very quickly formed a genuine liking for each another, in the beginning Robert was completely opposed to the whole idea of Ronnie's immigration. He called me for a meeting, motioned me to a chair and then said nothing for what seemed about ten minutes. This, of course, was one of Robert's strategies which used to upset people on the other end of the phone rather a lot. Very often he would be playing chess at the time, as he was now, but not being able to see him they didn't know that and it did have a rather intimidating effect. Normally I had no difficulty simply waiting it out, but this time his silence continued for so long, and since I also suspected what he wanted to talk to me about, I would just as soon not have been there in the first place.

Just as I thought I might get up and walk out he fixed me with that direct gaze of his and started to lecture me: 'You may have known this man for the past thirty years, but do you really know him? Once he comes out, if it doesn't work out you'll have the devil's own time getting rid of him. You are enjoying your life, Ethnée. You've plenty of friends. You love the stud, the horses. You're busy. You don't have enough time for yourself now. You'll have even less then. Have you any idea what you are doing?' Robert was too controlled to storm, but obviously he was rather upset.

While I was usually in complete agreement with Robert's judgement, in this instance we were not in tune. I found it hard to refute his argument, perhaps because I still had doubts myself, but I didn't like being told what and what not to do and consequently I felt the stubborn part of my nature emerging. I clenched my jaw in the way I have when I need to stop myself saying something I know I'll regret, and instead of getting angry tried to breathe deeply and count to ten. We came out of the meeting still friends.

It was only a few days later that an old riding injury of mine resurfaced. Twenty-five years before, I had damaged my neck rather badly while riding a horse with a reputation for refusing jumps. I had been determined that this time it would not stop … with the result that we took off too soon at a white gate. The horse fell and I came off, landing heavily and ploughing along the ground face-down. Although I didn't actually break any bones, I did fracture two vertebrae in my neck and my face was an awful mess. My neck was re-set under anaesthetic and my spine was stretched by pulling on my ankles. When I awoke to the dark blue bruises ringing my ankles I was heartily glad that it had all been done while I was under sedation. Although my neck has given me trouble ever since, I have been able to live with it by working around it. But this time, only days before Ronnie's arrival, I found my head locked into a very unpleasant downward position. I had visions of meeting Ronnie at the airport looking like a hunchback. I called our trainer Ted Martinovich, a chiropractor who worked on both people and horses. In no time he had clicked it back into place and I was, at least temporarily, both mobile and out of pain.

I was in a restless state of mind. Although I had been well-settled into my cottage for the past five years, now it seemed I had to fuss around moving a rug here, a picture there. But why was I doing this? Hadn't Ronnie said: 'I might as well bring out all my stuff. I've a lot of pictures and, of course, the carpet from Chikupi. Remember that?'

Did I remember the Chikupi carpet, the huge oriental rug patterned in brilliant shades of turquoise, reds and blues that was one of Erica's heirlooms? How could I forget? It had covered the living room floor at Chikupi and I wouldn't mind betting it has seen more parties and goings-on than any carpet before or since…

It is the night following yet another successful Lusaka Show. We're back at Chikupi celebrating as usual. Dinner is over and in the dining room the team of African servants, dressed in their crisp white kanzus with the red cummerbunds and red fez with black tassels swinging, are blowing out the candles in the tall candelabras, clearing away the china and crystal and polishing the beautiful inlaid table. Erica has departed for her bath and I snuggle into a corner of the comfortable old sofa in the living room, listening to the ice tinkling in people's glasses, the laughter, the chatter. Several eminent people including two well-known show judges are staying here tonight. The laughter dies down as Erica makes her entrance. She is dressed in a flowing floral housecoat. Laughing happily at the gazes fixed on her, she sinks gracefully onto the carpet. A whisky materialises and she stretches out her hand to take the glass, snuggles her knees up to her chin and proceeds to hold the floor.

Yes, I certainly remembered the Chikupi carpet. Would it fit here I wondered? But on the phone Ronnie's voice had continued, 'And there's a huge canteen of silver cutlery…glasses…all the crockery…do you think you can find a home for it all?' I remembered his multi-drawer cutlery canteen, too. Perhaps I would be better off making space rather than rearranging.

The words 'silver, crockery' kept repeating themselves. Silver, crockery. Silver crockery… They entered my conscious mind. It suddenly occurred to me that I hadn't set a table properly for years. My lifestyle being what it was, I tended to grab a quick sandwich when I felt hungry. Visions of Ronnie and Erica's table at Chikupi unfurled in front of me…

I called my friend. 'Joy, help!' She heard the panic in my voice. Doubtless she thought Ronnie had changed his mind and wasn't coming after all. Over the past few months, the word had got around that a tall handsome ex-cavalry officer was coming to stay, and at least three of my widowed friends had asked me if he had a brother. But for now the problem was a little more urgent.

'I need to do some practice dinners. Do you have any simple, *really* easy, cook books?' My mind recalled the careful coils of butter sitting in little crystal dishes of water. 'And I need to be able to do something with the butter! And napkins and flowers. In other words, I need a crash course in table setting. Please help!'

Bless her heart. Over the next few days we practised nearly every night. We folded napkins into contorted shapes that stood up unaided. We chose a colour scheme for the table which meant unearthing the dreaded iron—ironing is a close second to cooking in my list of dislikes—and ironing a tablecloth. I learned to make the butter look different, not perfect, alas, but definitely different, and the table was finally set with candles and a fresh posy of flowers as a centrepiece. We even decided on the dish for the evening of Ronnie's arrival. It was to be a chicken casserole that we would make together the day before. All I had to do was reheat it on the night. It may not have met the standards of Africa, but the result was attractive and welcoming and we were quite pleased with ourselves.

In his last phone call, Ronnie had said, 'Why don't we have

lunch when I arrive? A place of your choice…' I thought it a first-class idea. He was arriving around midday. It would allay any nervousness on both our parts and put off the moment when I might actually have to make something to eat. So I thought Kings Park might be a good place to go. It wasn't far from the airport and a wonderful vantage place from which to gain a first sight of Perth. Bearing in mind that he was six feet and six inches, it would also give him a chance to stretch his legs after the flight and before the trip to Heytesbury.

At last it was Friday, 6th October, 1978. Standing for what seemed like ages behind the barrier at Perth Airport, I spotted a tall slim man in a soft brown felt hat, sports jacket and cavalry twill trousers, manipulating his trolley through the terminal doors. My heart gave a little bump and I knew immediately that it was Ronnie. Suddenly he was in front of me. At five feet, I was a long way down but we managed a hug. I needn't have worried that we might be awkward with each other. We had so much in common we started talking immediately: Erica, other friends, the wildlife, our animals, horses…so much to say. The years fell away, and our original lunching plans seemed so unimportant that they just dissolved. I wanted to go straight to Heytesbury, to waste no time in showing him the stud and the cottage.

The cottage. I have to admit I was just a little nervous about showing him the cottage. After all, after the stately home in Dumfriesshire of which he had become laird on his father's death, what would he think of the small cottage that had originally been trucked onto the property for use as a weekender by Robert and Janet? Even before he came I had emphasised that it was a really 'very small cottage', now I mentioned it yet again. How would he adjust I wondered aloud.

'I shouldn't worry about it,' he had answered in the no-fuss way I was to come to know so well. 'I'm sure it will be fine.'

As it turned out, it was. But my nervousness reasserted itself in a fresh direction. We had been neighbours and friends for so many years. How were we to make the transition to being partners? Lovers? By the time evening arrived, this was increasingly on my mind and, I didn't mind betting, on Ronnie's too. Although we had been talking and laughing with each other for the past few hours as if the years between had never been, nevertheless there was now a rather dramatic and sudden change in our relationship. It was not so much that we felt awkward with each other, it was more an almost intangible air of uncertainty mixed with anticipation.

That night, we both changed for dinner as indeed we were to do all our married lives, but on this first occasion, we were both overly formal. Ronnie wore his regimental tie—blue-green with a silver stripe, which would later become our racing colours—while I had decided on a long blue dress of which I was rather fond.

Fortunately, earlier that day I had had enough foresight to transfer the setting to the table from the cupboard in which Joy and I had placed it the day before. The tablecloth was already ironed, the napkins had retained their intricate shapes and all I had to do was trim the butter and top it with a piece of parsley, polish two champagne glasses until they squeaked and hunt down matches for the candles. It was spring. I love flowers and the cottage was filled with bright pots and vases of the bush flowers I had picked from the slopes of the ranges early that morning. One of these I had placed on the table. Even to my critical eye, it looked inviting. Waiting in the fridge as planned was the perfectly browned casserole that needed only to be reheated in its dish. No pots and pans or risk of burning and spoiling were to mar the romance of the occasion.

And romantic it was. Ronnie popped the cork of the champagne. We toasted each other. The casserole was really quite delicious. The tie didn't last long. Nor did our first-night nerves.

❖

The decision to marry was mutual and almost instant. It just seemed right and quite natural. The attraction I had always felt for Ronnie, and evidently he for me, was as vibrant as ever. Neither of us had any impediments; we were both free to do as we wished. Or so we thought.

Our announcement of intention to wed wasn't greeted particularly enthusiastically by either Robert or his children. In fact, nine-year-old Catherine, always direct as I am myself, was quite open.

'Why do you want to get married?' she asked Ronnie in genuine bewilderment.

'Because it seems right. Once we are married, what will you call me?'

'The same as now. I'll still call you Ethnée's friend,' retorted Catherine rather grandly.

Why the hurry? Having made up our minds, there seemed no point in waiting. But there was another influencing factor in that my stepbrother Roualeyn Gordon Cumming and his wife Josephine happened to be visiting Perth from their home in Rhodesia at that time. They were planning to immigrate in the near future, but meanwhile their visa had expired and they had to leave within the next few days. I was fond of them both and, what's more, they had been to all my weddings. So although we had planned to get married in the garden, we discovered that such a ceremony would require three weeks' notice so we discarded that option and decided on a registry-office wedding.

The only hitch was neither of us could find the death certificates of our previous spouses, neither Erica's nor that of Charles. In the end, Ronnie had to send a cable to his solicitor in Lusaka, and I finally found a yellowing newspaper clipping which described Charles' death from the lethal bee stings he had received at Chobe and we were told that this would serve instead of a death certificate.

For both Ronnie and I this was to be a fourth marriage. Before his marriage to Erica and after his divorce from his first wife, Ronnie had been married to an Englishwoman called Connie, whom he had met in Egypt where her commander husband had been killed. She was apparently very beautiful, and consequently sought after, but it was into Ronnie's arms that she, literally, threw herself. In no time he married her, but it was another case of the proverbial 'marry in haste, repent at leisure' and time together was a ceaseless round of tears and trauma. On their return to England after two years in Buenos Aires as military attaché, Ronnie parted from her very thankfully.

But that wasn't the end of it. According to many accounts, Connie was a vindictive woman and Ronnie was not to cut himself free quite so easily. When he finally asked her for a divorce to marry Erica, she flew out to Lusaka with a lawyer and a support network of friends. The battle that followed would become the most talked-about divorce proceeding in the whole of Zambia. It went on day after day with the courtroom filled to capacity. It was a bad time for Ronnie…and extremely expensive. Finally he was ordered to pay her a small annual sum of money which he cleverly paid into an account with his solicitor in Lusaka. She was unable to take it out of the country because of the exchange regulations in place at that time. He told her she would have to visit Zambia in order to spend it…which of course she never did. When she died in England many years later, Ronnie gave the money, which in the interim had accumulated quite handsomely, to the Wild Life Society in Zambia.

Ronnie found hate a completely foreign emotion. He once told me he had never hated anybody, excepting perhaps this 'terrible woman'. One day in London many years later Ronnie met up with Connie's son by a previous marriage. Somewhat to Ronnie's surprise, this young man actually apologised to Ronnie for his mother. He said she had been most unfair and that he was sorry Ronnie had had so much trouble. It seemed he was under no illusion as to how malicious and spiteful his mother could be and it turned out that she had caused a great deal of unhappiness within the family. Ronnie must have gone through hell during that marriage but he said very little about it.

On Monday morning, exactly a month after Ronnie's arrival, we arrived at the registry office in St. George's Terrace which I have to say was as unexceptional as such places usually are. That is, aside from the registrar himself who was a tiny man who pirouetted throughout the ceremony with short, but rather high, steps. Ronnie was to name him 'our little ballet dancer'. Although Robert had refused to attend saying that he hated 'that sort of family affair' Janet was good enough to come to our little ceremony. And of course Roualeyn and Josephine were there as witnesses.

And so, for all the ordinariness of the surroundings, Ronnie and myself were married in a shaft of sunlight that was to last for twenty-one golden years.

Ronnie had planned our celebration lunch at the Parmelia Hotel in the centre of the city. It was a truly lavish affair held around a table spilling over with the most exquisite red roses. The staff spoiled us with attention. The food was delicious. I could have gone on enjoying it forever. Although she is now living in Washington D.C., Josephine says laughingly to this day, 'I've never known anyone eat so much. I think you both knew that you weren't going to cook for each other!'

The almost rarified feeling of happiness followed us home to our cottage. There, piled up beside the front door were more beautiful flowers, some sent by Heytesbury's head office, others from our friends. Just as we were bending to pick these up, there was the sound of wheels on the gravel behind us.

The taxi stopped outside the front door and there was another gorgeous bouquet of red roses with a card from Robert wishing us great happiness. I appreciated that gesture more than he would ever know.

Over the years, we rarely cooked for one another, but we certainly made each other extremely happy. That's when I realised that it is possible to love deeply more than once.

CHAPTER TEN

The 'Fabulous Thoroughbred Stud'

It was 1978 and Robert was well on his way to putting in place the strategies that would earn him the title of Australia's richest man. Outside Heytesbury Stud, the Holmes à Court empire was growing on a tide of acquisitions: among them majority holdings in two leading Western Australian companies, Albany Woollen Mills and Bell Resources. The stud was Robert's 'time out', his hobby, his love and his chance for relaxation. He took an intense interest in every aspect of its progress. And this, too, was developing at quite a pace. The original property had been expanded by the addition of land holdings on its perimeters: land belonging to the old dairy on its southern boundary and further acreage across the road which was to become Little Heytesbury and used mainly for the rearing of yearlings.

Ronnie had first met both my sons when they were young boys in the show ring at the Lusaka Agricultural Show. On that occasion, Robert was riding Toshey—a pony Erica had lent him to compete in the showring—and as I have said he handled the plucky little horse so well and with such confidence that she later gave it to him. But now Ronnie and Robert were to meet as adults.

Because of Robert's initial reaction to Ronnie's decision to immigrate, I was a little nervous about this first meeting. Thirty years had passed. It was now a question of two men meeting each other and, probably for both, it would seem like the first time.

But I needn't have worried. They were two quiet and clever men who were well able to respect the qualities they saw in each other. They took to each other immediately. Of Robert, Ronnie said, 'Little did I think when I saw that confident kid in the show ring that Robert would achieve what he has. I was completely unaware of the genius that Robert would become...' And later he was to re-capture those first impressions in his memoirs, 'Robert had patrician looks and a fine physique. He had already made a name for himself as an up and coming businessman. Janet, his wife, had brains and beauty. With two such gifted parents it was not surprising that their four offspring showed promise.'

Robert, on the other hand, showed his respect in a different way. Very little time had elapsed since our wedding day when he called me.

'Ethnée...I want you to draw up a plan for a new house. Any plan you want. Choose yourselves a spot anywhere you like on the stud and I'll build you whatever you want.'

At first we were both delighted. We drove all over the stud and chose three or four sites, all of which would be perfect for our new house. I had plans drawn up and each evening we would sit down with a drink and move a paper wall here or there.

One evening, we were sitting out under the pergola on the back patio, scotch and ice to hand, listening to the water splashing into the little pond I had been constructing the day I had first heard Ronnie was thinking of coming to Western Australia. My proud peacock followed closely by his two faithful guinea fowl passed jerkily by on their evening rounds. It had been a busy day as usual and we were enjoying the silence that comes with the best companionship.

Ronnie turned to me. In that quietly considered way of his, he said, 'My darling, why do we want to move? Are we sure we want to move? This is where your memories are... It's where your Simon stayed and worked on his film. All our pictures and furniture fit. Your friends from Africa have stayed here. The convenience of the foaling-down yards, this beautiful evening light shafting through the gums... I wonder... Could anything else be more perfect? Would it be possible for anywhere else to have more memories? Be more comfortable? Really?'

I looked at him with some surprise remembering the uncertainties I had had about what his reaction to my little cottage would be and I realised I had underestimated him. I realised, too, that he was right. The real values in life and in our lives, all that was important to us both, lay right here. We had it all. There was no need to move.

And so this is where we elected to stay...and where I still live another twenty-five years later.

Paul, the youngest of Robert and Janet's children, is five years of age. He is already a fine young fellow, showing signs of the young man he will become.

Today he is going to play the violin for us, a special concert for Ronnie and I. The three of us are standing in the living room of the cottage.

'Where shall we sit?' I ask.

'Anywhere, anywhere. Oh, over there is okay.' He waves us to the armchairs.

Once we are settled, he stands in the centre of the room, brings the instrument up to his chin, flexes the arm that holds the bow. Carefully, he places one foot slightly in front of the other. Frowns. The balance is not right. Shifts feet again, and then again. At last he is satisfied. His attitude is perfect. The confidence he projects reminds me of Robert at a similar age.

He plays a few bars of *Rattle Rattle Dump Truck*. We watch and listen. The concert is brief. The tune ends too soon but I know I will remember it always.

'That's it,' he says uncertainly.

Ronnie gets up, bends down to the young master. Congratulates him. Shakes his hand.

Almost every weekend Robert and Janet and the youngsters continued to drive down to the property. The four children were growing up. Catherine and Peter were eight and nine respectively, Paul and Simon were five and six years of age. The stud was not only an ideal background for their fierce outpouring of energy, but an excellent contrast and antidote to city life. Schoolfriends often arrived with them, leaping out of the car, their faces almost split in two with smiles of excitement and anticipation. Importantly, too, the daily business of the stud—the checking of the mares, the matings, the births—as well as the farmland environment, became a natural part of their lives.

Ronnie had never wanted children of his own. He enjoyed his life of hunting and fishing too much to see himself as a conventional father and each of his four wives already had their own children. Here, however, he took on the role of grandfather, relishing his relationship with Robert and Janet's young family, and they in turn enjoyed the stories of his adventures through Burma, India, Egypt and southern Africa. He was not only a great raconteur, but he had taken part in real-life adventures. It was real boys' and girls' own stuff and for one tale he had the evidence in the form of a piece of camouflage-coloured silk torn from his parachute. No matter how often he retold the story of the time he was parachuted into Burma, no matter how often he brought out the scrap of greenish silk, they were unable to take their eyes from his face except to gaze on and gently touch the flimsy piece of material.

Although he got on well with them all, perhaps it was because Paul, being the youngest, was at the stud more than the others that they formed the closest bond. As the years went on they became closer still. One of our evening talks about trout fishing must have stirred Ronnie's nostalgia for his sparkling Scottish streams because the following day he decided to teach Paul to fly fish.

From then on it was regular practice on the back lawn: little talk and lots of action, the lines snaking out over and beyond my pond to the gum trees. The day Ronnie felt that Paul was ready for some real action, they took off together, rods slung across their shoulders. We had stocked one of the largest dams with fingerlings and the trout were just large enough to provide both sport and a tasty feed.

Evidently Paul never forgot his lessons because some time later, during a year spent on the Oppenheimer's beautiful game farm in South Africa after completing school, he won a prize for catching the largest trout. To prove this was just not another fishing yarn, he had the fish stuffed and today it hangs on the wall in his Claremont house…proof of the big fish that didn't get away.

But camping was just as much fun as fishing. By this time there were quite a number of campsites on the property, but there was no doubt that the favourite was the island in the middle of the house dam. This was a 'by invitation only' site, absolutely *verboten* to friends, family and any adult who had not been specifically invited. The island was a true island, separated from the shoreline by fifteen metres all the way round, making it inaccessible to anyone without a dingy.

As 'official photographer', one day I was issued with an invitation to visit this precious camp. I was ferried over to the island in the small dingy with a full foot of water in the bottom of the boat, which constantly needed bailing out, and proudly shown the retreat. My job was to take pictures of island activities, which resulted in photographs I still treasure today. Peter was snapped reclining rather self-consciously on a hammock while his best friend Ben, and Ben's brother Andrew, were hard at work. Ben was cooking on a precariously balanced stone 'stove' and Andrew was washing the utensils as they were used. It was all very organised and gave us some foresight into Peter's leadership skills which he would put to good use in his early adult life.

Peter and Ben spent a considerable amount of holiday time on this little island. In addition to the little dinghy, they had a raft for carrying equipment back and forth. Although this was no Kontiki, put together as it was with empty forty-four gallon drums and old planks held in place with rather a lot of wire, it did provide a very useful purpose. When it was not conveying tents, cooking utensils, tools, firewood and rations back and forth, it doubled as an excellent diving board.

I must have passed muster not only as an official photographer but also as a non-interfering adult because it was not long before I was invited to another camp. This one was situated among the tall trees on the top of a hill and this time I was 'allowed' to bring Ronnie. We were greeted with some ceremony and led to our rather wobbly log seats, rather low for Ronnie's height, which caused him to relate one of his college experiences: 'I was hopeless at running. I seemed to remain in the same spot forever. But I always won the high jump because I simply stepped over!' The occasion was marked by afternoon tea brewed in a billycan and Peter was the tea-maker. He hung over the fire willing the billy to boil while Paul and Ben rustled around in the bush busily keeping up with the demand for a steady supply of firewood. Although it was a long time in the making, the tea served up in the enamel mugs was delicious.

That evening, Ronnie and I were having dinner when the boys arrived at the back door of the cottage with three dead rabbits hanging rather bloodily from their hands. They were proud at having shot them with Peter's small air gun and wanted Ronnie to show them how to skin the carcases. Ronnie immediately slipped back in time to his own childhood days in Scotland when he went out on hunting forays with the gamekeeper. With a show of glee that surpassed even that of the lads, he removed his special skinning knife from its leather sheath and took the carcases into the garden. The lesson proceeded, but the boys managed to hover half in, half out of the doorway, part of their attention on what Ronnie was doing, the balance firmly fixed on a bowl of dessert which sat waiting to be consumed on the dining room sideboard. While Ronnie's cooking skills could not be rated as one of his major strengths, he could certainly make the most marvellous desserts.

For this particular evening he had prepared one of my favourites: a delicious concoction of peaches and whipped cream flavoured with lemon. He had made it the night before, so that the flavours would combine for a melt-on-the-tongue experience. This was the dish that attracted the boys' glances, so once the rabbits were skinned and one offered to us as a present for the following night's dinner, how could we not return the compliment? The three dirty boys washed their hands quickly and superficially and tucked somewhat more enthusiastically into the dessert. Did we blink? I don't remember. Certainly it seemed only a fraction of a second later that the bowl was empty and only a second after that that the boys trudged off into the night, waving us a happy goodbye and swinging their bounty by its skinned heels.

Given their father's love of horses and natural riding abilities, it was inevitable that his children should follow his lead. And although Janet didn't share Robert's enthusiasm for horses, she certainly didn't discourage the children from learning to ride. From a very early age I gave each child lessons: either sitting in front of me or by leading them on Impion. Once they were older they progressed to owning their own ponies and attending pony club camps.

Although Peter's first riding pony was a grey with a chestnut patch on one side, called Patch, technically-speaking it was not his first horse as he had been given a foal by Ernest Lee-Steere some years before. Robert and Ernest used to practise polo from time to time on Ernest's farm and when the mare that Robert rode produced a foal, Ernest gave it to Peter of whom he was very fond and whom he used to call his 'godson'. But the maturity of Peter and the foal didn't progress at the same pace and it wasn't until Patch came along that Peter's riding days really started. He was keen on looking the part right from the start and insisted on having a 'cowboy' saddle for Patch; a rather fancy chased leather affair with a bridle to match with, of course, similarly flamboyant attire for himself.

Paul's first horse was a broodmare called Snudge, the first of many he was to own over the years. Perhaps it was this first mare, out of which the racehorse Vegas Vixen was bred, that started Paul's interest in bloodlines. Vegas Vixen turned out to be a prolific winner and her progeny is still seen on the track. Paul loved horses and riding right from the very beginning, showing the same affinity for the sport as his father and I.

Catherine, too, took to riding with the same enthusiasm I had so many years ago. She became such a skilled rider that when she reached her teens, Robert gave her a beautiful chestnut thoroughbred called Bobby which she used to ride regularly in all the Pony Club events within reach. Although it meant rising very early in the morning, Ronnie and I made a point of attending as many of these as possible. We enjoyed the outings in themselves, but Catherine was our focus. We knew she was good and that she would go far, and we wanted her to know that she had our full support.

Simon was the only one of the four who didn't take to riding quite as fervently as the others. Instead, his passion was for computers. From the age of eight, he was never happier than when he was sitting in front of a computer screen. A secretary I had in the office for some time had a son, Wade, a boy much older than Simon, and they spent an enormous amount of time on their machines. But perhaps Simon's computer was all the more treasured because he bought it himself. To raise the money he offered to paint the post-and-rail fences of the foaling-down paddocks. Before he started, the $1 per panel recompense he had been promised had seemed quite generous, but when at the end of a long day he had only 20 panels painted and he found the exercise as boring as trotting over endless cavalettis, he could see that it was going to take both time and perseverance to buy his dream machine. What's more, the white paint was turpentine-based and took a long time to remove from places it shouldn't have been. But buy a computer he eventually did...and as it turned out, a noteworthy technological career was to become his future.

All the children had their own broodmare or a racehorse, sometimes both. They would purchase a horse from Robert from whom they borrowed the money for this purchase at what seemed to them to be an extortionate eight per cent interest. Of course, the foals then became the property of the youngsters and they had the option of selling the yearlings at the annual sales. Catherine was always the lucky one in that her mares foaled regularly and she always seemed to get the best prices for the offspring. Right from the start Robert instilled a business sense into his children. If they had not borrowed the money at the outset, they would not have been able to buy a mare in the first place. They learned the value of a broodmare if she produced a foal that sold well at the yearling sales thus returning a profit on their original investment.

Of course, the foaling season provided the children with natural high points of excitement. Since mares tend to foal at night and since the ultimate thrill was to witness a mare of their own giving birth, it was a particularly special treat to be allowed to take their sleeping bags to the 'look-out' area, a small tower block that had been built on to the end of the stable complex overlooking the foaling-down yards.

Sometimes a friend would be invited to witness the foaling down. And usually the larder would be raided ahead of time for a 'midnight feast' which would inevitably be devoured well before the appointed time. But there were many occasions when the elation of the whole adventure just proved too much for the children and they fell asleep. Words cannot describe the looks on their faces when they awoke in the dawn only to find out, that for all the planning and anticipation, they had slept right through the event.

Meanwhile, I needn't have worried that Ronnie would not have enough to do. He took a great deal of pride in keeping fit— possibly one reason that he lived to the grand age of ninety-four—and it didn't take him long to form the habit of going for a brisk walk over the stud for miles every morning. He would come swinging home from these forays with a happy smile on his face and the colour high in his cheeks.

He was determined to exercise and spent a lot of time outside helping with the planting of yet more trees. Another chore he undertook was to mow the grass of the verges. This involved manipulating an unusually heavy mower around the stud. Janet adored Ronnie anyway, but since much of the design of stud and the plantings were due to her efforts, when she saw the neat verges, she was particularly delighted. 'That does look nice,' she said approvingly and shortly after that I noticed that there were two gardeners with mowers helping him with the job. But it wasn't so nice the day the mower caught fire. In the heat of summer the flames spread quickly. I was in the office at the time and heard sounds of commotion not far away. Apparently, the machine had hit a rock and burst into flames which caught the overhanging leaves of the volatile gum trees. Although we had our own fire-fighting equipment, it was a relief when the local fire brigade arrived to augment our efforts.

Water is becoming increasingly precious all around the world, not least in Western Australia. By now we had eight dams on the stud from which most of the pastures were irrigated and we used as little scheme water as possible. However not all the trees on the property had reached the majestic height of the eucalyptus grandus behind my cottage and the younger saplings still needed water, so this was another job Ronnie took on. He made a saucer around each of the hundreds of trees and he would walk alongside the water cart making sure each tree received just enough.

Or sometimes I would look out my window to see Ronnie driving the cart while my grandson Simon ran alongside moving the hose.

These were happy years. I was in love with my life: my husband, my sons and my grandchildren, my beloved animals and the stud. My life had richness and variety. I felt that there was nothing else I wanted, that life couldn't get any better than this.

But as it turned out, it could. And did.

CHAPTER ELEVEN

Heytesbury Heads for the Stars

By the late Seventies, Western Australia still had no national record in racing. We were well behind much of the rest of the Australian field in this regard. But we were coming to the end of an era when everyone or everything promising automatically disappeared 'interstate' or overseas. Western Australia was just starting to move forward.

Heytesbury, too, was on the move. It was not only starting to show the promise it was to fulfill as a premier stud, but fast becoming a prime property in every respect. Robert never stopped innovating and improving. Most of our staff were now fully qualified, although from time to time we still had people working for experience and during the busy foaling-down season or when preparing yearlings for sale, we started the seasonal practice of employing extra staff. During the stud season, we employed a vet—usually from America, but one year from Ireland—who resided at Heytesbury for the duration and, of course, at all times of the year we had a vet on call. For many years the foaling-down yards had been floodlit, and we were now able to instigate a watch twenty-four hours a day from the 'watch-tower' at the end of the stable complex overlooking the yards. The white post-and-rail fences that had so quickly replaced the barbed wire a few years earlier were in the process of being painted black, an unusual idea Ronnie and myself had brought back from one rather stunning Kentucky stud.

Another idea we had picked up on our various forays through international studs was a blueprint of the Irish National Stud serving barn, far and away the most impressive of its kind that we had come across. This unique building was octagonal in shape, which together with its lofty ceiling-less roof, added to its spaciousness. I took my usual scads of measurements and photographs to discuss with Robert once we were back home. Typically he wasted no time in saying, 'Well, if it's good enough for Ireland, it should be good enough for Heytesbury. Let's build it.' A little later, we did, along with a fully-equipped veterinary complex.

It was always Robert's belief, probably as a result of the knowledge he had absorbed during his agricultural education at Massey University in New Zealand, that two of the most important aspects of stud breeding are 'climate and the right pasture'. He was convinced that these two features were key in the breeding of 'outstanding horses'. And it was outstanding horses that he was determined to produce at Heytesbury.

For Robert, 'the right pasture' was a case of controlling the weed population and experimenting with varieties of pasture that would yield a balance between providing lush feed and continuing to grow through the dry summers. As with everything my son did, as little as possible was left to chance, and an officer from the Agricultural Department was called in to take soil samples for scientific analysis before the paddocks were seeded with the most appropriate pasture.

Weed control was more difficult. We have never been fond of chemicals and whenever possible looked for other solutions. First to undergo an experiment was Boyd Road Paddock (as with Rosebush Paddock, all our paddocks were named) which had a particularly bad growth of dock. This had been introduced with some hay we had brought in and was proving almost impossible to eradicate.

Again, Robert called in an expert and we were advised to proceed with a three-year plan. For the first year, it was suggested that we plant wheat—which we thought somewhat strange since Heyetesbury is certainly not in a wheat-producing part of the country. We must have been the only farm for hectares around with a wheat crop. In the fullness of time, this was cut, stooked, and finally cleared away, a comparatively easy operation, save for a shock encounter with a long snake that writhed out of an armful of wheat one of the girls had gathered up. The second year, we were to run sheep to graze on the stubble. This involved enclosing the paddock with ringlock, since almost all the stud's paddocks and yards were fenced with post-and-rail, ideal for horses, useless for sheep. We had barely finished this task when in came the sheep on the back of a ute, complete with sheepdog. This was another instance when I would have liked to have had a camera handy. It was a little nugget of a scene that was pure Australian and I remember thinking that it compared well with the Africa I had left behind. The following year led to the final phase of the experiment: seeding the paddock with pasture once again. To this day, there has been no reappearance of dock in Boyd Road Paddock. And it was all done without the use of chemicals.

While Heytesbury is now almost totally weed-free, Paterson's Curse—Salvation Jane in South Australia—is another unwanted weed that occasionally reappears on the property. This is the weed that is also called a wildflower. Tourists love it because it turns the paddocks and hilly slopes into vistas of dramatic purple in the spring. When I first arrived, somewhat to my shame in the light of what I now know, I thought the plant so pretty that I would pick the flowers and display them in a vase. Now, whenever I come across it, I pull it out by hand and burn it. But it does continue to surprise me that what is unwanted in one State is accepted in another. We call the plant Paterson's Curse because it stops anything else from growing, while the

South Australians welcome it as Salvation Jane for its (very limited) feed value.

Another regular chore that ranks alongside collecting sticks and split bark from around the bases of the eucalypts is collecting the rocks that rise to the paddock surfaces from time to time. In response to a suggestion made half in jest after lunch one day, some visitors decided to go rock-collecting. It was quite a sight to see a phalanx of people vigilantly scanning the ground for small rocks that they dropped into their progressively bulging sacks. We've had other visitors who took part in gathering lupins—another pest for studs—for burning. I suppose they regarded it as a good way to work off a big meal.

But while we managed to control rocks and weeds, one chore that increased in proportion to the growth of the eucalypts was the collecting of sticks and shed bark from beneath the trees. To take care of this, a 'stick run' was instigated. This involves driving around the stud and picking up sticks from the verges beneath the trees and generally clearing any other rubbish. Because this 'chore' incorporates the use of a vehicle, it is perennially popular with the young. There is always a child who has just learned to drive, and the small Boulins tractor or the 'mule', with top speeds of around 15 kilometres an hour, are a comparatively safe way to improve their skills and earn some pocket money at the same time.

I remember two youngsters enjoying this chore so much that they would sing their way through the job, completely off-key and at the tops of their voices, oblivious to the fact that on a still day they could be heard for miles around. Disaster struck this pair one afternoon when the 'driver' became so over-exhilarated she took the corner rather too quickly, turning the vehicle onto its side. Fortunately, neither child was hurt, just shamefaced. But it was a considerable time before these two, or in fact any other child living on the stud, were allowed to resume the responsibility of being in charge of a vehicle. Terrible punishment.

It was not only this close attention to running a tight ship that worked in the stud's favour. In addition, Robert had an uncannily retentive memory which had contributed to his reputation as a 'master of bloodlines'. Few people could match him on the subject. He had a faultless recollection of which stallion had been put to which mare, at what date and with what result. This naturally led, in turn, to his expertise in breeding.

At this time the press was full of the different methods Robert was trialing in order to produce a 'super horse'. They maintained he was not only after a champion, but one with unparalleled bloodlines that would revolutionise the breeding industry. Although some of this was media hype, in other ways, they were quite right. As with his unquestionable talent for picking stocks and shares, his lateral mind worked to the same advantage in the horse breeding industry. He weighed up the pros and cons of conformation, performance and stamina, to which mix he added a dose of pure instinct. As the late Tommy Smith, one of Sydney's leading horse trainers, said, Robert had the best eye 'for a horse of anyone I have ever met'.

Since 1979, Robert had been scouring the world for mares bearing the richest thoroughbred blood. At one point he had something like eighteen mares waiting in the United States— gathered from countries including Germany, Holland and England—to be transported to Australia. By the end of that year, we had fifty broodmares on the property and another thirty fillies in training around the world.

Although Heytesbury now had two excellent stallions, Silver Knight and Haulpak, Robert saw a need for at least one more. It was in Victoria that he finally found a stallion whose stakes record and breeding appealed to him. This was Family of Man, another record stakes winner whose list of triumphs included the Australian Derby and the W.S. Cox Plate. At the time, he was in the care of well-known trainer George Hanlon who had him in work in Melbourne. It was not long before he and Robert had reached an agreement and the big bay, with the enormous blaze that could be seen way off down the racetrack, was loaded on to a float for his trip across the country to his new home.

It was only three weeks after this that Robert made another purchase. Family of Man had barely settled into his new home, when Pago Pago arrived to join the stud. Many of the horses that made their way to Heytesbury had colourful backgrounds and Pago Pago was no exception.

He was foaled in Australia on 10 October 1960 and started his life over East. He made his mark in his home country as a two-year-old by winning the highly-esteemed Golden Slipper in 1963. But this was the era when we allowed so many of our promising players, in a range of careers and disciplines, to leave the country. The winning margin and style that Pago Pago displayed on this occasion were such that he was fated to follow the trend. He was immediately purchased by Americans and shipped to the States. On this trip, so the story goes, a film was shown which portrayed horses neighing in some distress, which heard in concert can be both noisy and alarming. When the discordant sounds floated down to Pago in the hold, somewhat to the amusement of his strapper, he responded enthusiastically in sympathy. He landed safely in America where he was to stand for the next 17 seasons—first at the famous Clairborne Stud in Kentucky and later in Florida.

Meanwhile, Robert had made a point of following the success stories of Pago's progeny. Over the years he became so impressed with the results of the offspring that he decided to purchase the sire. He tracked him down to a stud in Florida, where he arrived one day, cheque book in hand, and successfully proceeded to persuade the owners to sell. But by this time it was 1979 and Pago Pago was nineteen years of age which engendered a great deal of criticism of Robert's decision. Many people thought he was taking an unnecessary risk in buying a stallion of this age. But they forgot that Robert had vision! That vision, as usual, was to pay off.

But I must admit there was a short spell at the beginning of Pago's life with us when even I wondered whether Robert had made the right decision.

Originally Pago had traveled to America by sea, as in those days there was no air travel for horses. However, we decided to bring him back by air. By this time, air transport for horses had become quite sophisticated and he actually traveled surprisingly well and appeared quite relaxed back on the ground again for the short trip between Perth Airport and Heytesbury.

It was not until he arrived at his new home that he became unsettled and when we finally let him out of his stable into his lush paddock, he 'walked the fence', back and forth along the post-and-rail in such a frenzied manner that it became obvious he was not going to settle unless we did something to calm him. But what? It was Ronnie who came up with the answer. During his years with the 13th/18th Hussars in India, his cavalry regiment consisted of over six hundred horses, or 'walers' as the Aussie horses abroad were called. Whenever a new 'recruit' arrived and would not settle, a seasoned 'waler' was put in the paddock with him to help him to relax. It appeared such a simple solution, but apparently it always worked. We took a risk and put one of our old hacks in the paddock with Pago Pago. It worked like a charm and he settled almost immediately. But for many weeks to come, the hack had to be within sight.

However, the test of Robert's judgement was still to come. The stud season had already started and Pago was taken to the serving area. This was in the days before our serving barn had been built and the shady, rather idyllic pasture with the huge gum trees was altogether different from what the stallion had been used to in the United States. In fact, he was so busy taking stock of his surroundings that he was not at all interested in the mare. He seemed to be saying to himself, 'I'm not quite sure where I am…' We were beginning to think the mating would never happen, when suddenly out of the clear blue afternoon air came the 'laugh' of a kookaburra. A kookaburra is a rather large fat-looking Australian bird that is sometimes called a jackass because of its lengthy call which closely resembles hysterical laughter. We saw Pago's head snap around and, as he listened to what must have been an almost forgotten sound, his eyes cleared. You could almost hear him say, "That's right. Yes, I'm home…' whereupon he smartly remembered the reason for being introduced to the lovely young mare!

Once again, Robert was proved right. Pago continued to perform his stud duties until he reached the grand age of twenty-seven, at which point we decided to retire him. Just as has happened with Haulpak, his progeny is still prominent on the racetrack

We celebrated Pago's twenty-first birthday at the stud in 1981, which was an event that attracted both his fans and the media in their dozens. He received several cables of congratulations which were read out during the course of the afternoon, including one from the Blood Horse Society in Lexington, Kentucky, his adopted country. One thoughtful admirer brought along a huge home-baked carrot cake which we carefully offered to Pago who was not at all impressed. He stretched out his neck, and sniffed the cake a little and then quickly turned up his top lip as a horse is wont to do when there is a nasty smell! The staff were the winners, as they were able to enjoy a very tasty cake.

As the time approached for each stud season, it became a matter of routine for Robert and I to discuss the breeding programme which revolved around which mare would be put to which stallion—and why. Breeding, conformation and performance were all important considerations. For example, if a horse was 'slightly back at the knee' as was Silver Knight, he would be mated with a mare with the opposite trait. Good stud conditions offering the right atmosphere and environment were also vital ingredients. We made sure the stallions were free to roam in and out of their boxes. The mares grazed peacefully in the dappled sun with their friends nearby. Food, shelter and a good supply of kindly words and sliced carrots were their daily lot.

As I have said, Robert's aim was high and usually during these 'discussions' we would talk throughout the night, rarely stopping for either food or drink until the early hours of the next day.

Time and again Robert would stress the importance of speed as well as stamina, and part of his breeding strategy in the quest for his 'super horse' was to combine the bloodlines of both stayer and sprinter. Although he did this successfully many times, probably the most famous match was the mating in the 1978 season of the sprinter Brenta to stayer Silver Knight. On August 29, 1979, Brenta gave birth to a colt so black that there was no doubt in our minds we would name him Black Knight. I must admit I find it somewhat ironic that despite the fabulous prices we have paid over the years for what were undoubtedly top mares, it was Brenta—one of the foundation mares for which we had paid the royal sum of $600 back in 1971—that was to be the dam of the first WA-bred horse to win the nation's greatest race. When Black Knight won the Melbourne Cup in 1984 he produced a winning double: for both Heytesbury and Western Australia.

The breeding and environmental programme that produced Black Knight was only part of Robert's strategy. His longer-term planning provided the rest of the key. Black Knight showed promise from the start. He was initially placed in the hands of a Perth trainer, Roy Edwards, and won barrier trials from the beginning. But this did not suit Robert. He wanted a slower beginning and infinitely greater future for the horse.

Black Knight was transferred to the stables of Victorian distance trainer George Hanlon and a twelve-month programme was drawn up. Right from the start Black Knight was aimed at the winner's circle at Flemington for 1984. Over the next year, the horse was to cover over 20,000 kilometres under race conditions over 12 outings in the run up to the Cup. The media was excited. Headlines proclaimed the WA-owned horse a serious contender for the great race. Black Knight was fit and keen and ready to win.

If only we had known that! Because to tell the truth, no one actually thought he was going to win. On Cup Day that year, only Janet and her eldest son Peter were at Flemington racecourse. The other children were at school. Robert was in London on business. Ronnie and I were at a Cup luncheon at the Parmelia Hotel, our sights focused on a large screen. Black Knight's odds were 10/1 and I punted the grand sum of $2 to win. As far as I know neither Robert nor Janet had bets on Black Knight, although I believe the children placed $10 each way.

Flemington is a great place to be on Cup Day. The air of anticipation, the women's fabulous outfits and hats, the months of build-up for owners, trainers, punters—and milliners—alike are all released in the finale. But nothing seen at Flemington could beat the jubilation in the restaurant of the Parmelia the day Black Knight passed the winning post ahead of the field. There was a split-second of disbelieving silence before a huge cheer almost lifted the ceiling. This was the first time the race had been won by a horse from the West. Everyone was so excited. I've never been kissed and congratulated by so many strange men.

Robert, on the other hand, was asleep in his London hotel at the time. He was woken by Janet calling with the news from the Victoria Racing Club offices just after she had accepted the $23,000 Gold Cup trophy. I can imagine his quiet sense of satisfaction. Everything he believed in—his breeding programme, the environment and conditions he had produced for his horses—had come up trumps. Heytesbury had produced a winner of Australia's greatest race.

CHAPTER TWELVE

Winds of Change

Robert regarded George Hanlon as one of the greatest trainers in the country; the only difference of opinion between them that I can recall being over the stallion Family of Man. While George rated Family of Man as the best horse he had ever trained (*West Australian* 1999), to Robert he was somewhat of a disappointment. When Robert purchased him in 1979, he was one of Australia's leading stakes-winners with a sound 20 wins out of a total of 59 races. But although he served a full book of 38 mares in his first season at Heytesbury, he was not as well-known in the West as in the East and Robert thought he would do better back in Victoria.

For Robert thought was quickly translated into action with the result that Family of Man—accompanied by a rather nervous George Hanlon, who hated flying, and my grandson Peter—was flown to Heytesbury's beautiful Wallan Stud, some forty kilometres out of Melbourne. Among the undulating hills of the Victorian property, an equally strong book of mares was immediately produced by the announcement that the stallion would stand free of charge, with only a handling fee being levied. So instant and satisfying was the response, so commercially viable the progeny in their home state of Victoria, it proved once again that Robert had made the right commercial decision.

Meanwhile our Melbourne Cup winner Black Knight was

destined for a very different career. Robert decided to give him to the Victorian Police Force, depending on how he conducted himself during a three-month cadetship. Black Knight proved he had the right temperament and passed all his tests with flying colours. As a police horse he still gets to take part in the Melbourne Cup Parades and I wonder what he thinks when he hears and sees the people, smells the suspense and hears the roar of the crowd…

Sometimes Robert alerts me when he is coming down to Heytesbury; sometimes he arrives unannounced and unexpectedly. Like today. It is the middle of the week and a call comes through from the stud office. 'Let's get the horses saddled, Ethnée. It's time for a talk and a ride.'

We ride together slowly and easily as we have done all our lives. Robert is a beautiful rider. He sits his horse as if he were born in the saddle, as I suppose he almost was. Certainly, he was only a baby when I first placed him in the saddle in front of me and rode off into the Rhodesian bush. We climb into the ranges, discussing the water, the fencing, the 1986 stud season fast approaching. We pass the spring-fed dam nestled into the hills. If all else were to fail, this dam has enough capacity to irrigate the stud for the best part of a year. I tell Robert this is where I think we should try to sink a bore. We have tested for a bore lower on the property without success, but up here, the grass is lush and thick. I have a hunch there is plenty of water here. Robert appears to agree. He nods, but sometimes I wonder whether he is actually listening and agreeing or following his own line of thought.

Towards the top of the property, we stop and dismount. The air is keen and clear today and we can see right across the flats to the ocean. We absorb the peace, drinking in the sense of achievement we get from looking down on the past fifteen years' work. From this height, the stud is laid out before us like a toy farmyard. Directly in front are the thirty-one shelters and their adjoining paddocks which we use for weaning and the preparation of the yearlings. Neat rectangles run one beyond the other, all edged with the glossy fence railings that look like matchsticks, all with their shade trees towering over the miniaturised livestock. Other larger paddocks stretch to the south, another stable complex, the parade ground, the foaling down yards to the north. The shimmering leafy heads of the silver-barked eucalypts delineate the avenues. Closer to us, on the lower slopes, a herd of blond d'Aquitane cattle nod over their ceaseless grazing. Someone is moving broodmares from one paddock to another and the sound of their hooves, their neighing, lifts on the light breeze. Otherwise the silence is absolute and it's hard to believe that over two hundred and fifty head of the world's premier thoroughbred stock are tranquilly grazing or dozing in the pastures below.

I fancy I can see Silver Knight's silver-white coat. Haulpak and Pago are more difficult to spot.

Robert speaks with a short laugh: 'Remember when it was all barbed-wire and rubbish?' He turns to me. 'Ethnée, I've decided that the time is right to build another family weekender; this time it'll be a big weekend retreat, a *big* weekend cottage…'

The site they picked was the first slight rise into the ranges, about a kilometre in from the front gates. The house would nestle into the hills rising behind it, views from each room reaching out across the stud.

It was in the building of this house that Robert showed his love for the two countries that were so much part of him: the Africa of his birth, the Australia of his choice and successes. As the building progressed, it became clear this would be a house that encapsulated the essence of Africa with the characteristics of Australiana. A long eucalypt-edged avenue sweeps up to the house. The walls are of limestone, with wide verandahs on three sides supported by solid-looking gum poles. Wooden shutters flank the windows. To supervise the construction of the shingle roof, an artisan was flown out from Canada. Although it sits among the redgums, jarrah and blackbutt, growing out of the landscape like the Australian homestead that it is, those who have lived in Africa insist there is 'a bit of Africa in it'.

To one side of the house, fenced for safety, since small children are an important part of the Heytesbury family, is a swimming pool and entertaining area complete with barbecue facilities. West of this main house is a smaller duplicate. This is the guest house with four huge en-suite bedrooms off a central kitchen and living room. Attached to the guesthouse are a gymnasium and squash court and a little further away a tennis court, all of which are freely used by staff on the occasions when we have no guests. Available for use by staff, too, on certain days of the week is the swimming pool at the main house, a real boon after knocking-off time on scorching summer days.

Three years earlier, Robert and Janet had moved their city residence to a large house in Peppermint Grove. There the four children had been allowed to choose their bedrooms in order of seniority, with the result that Paul, the youngest, had ended up with the smallest room. With the building of the new house—this comparatively extravagant 'weekend cottage'—Robert announced that the order would be reversed. Paul was allowed to choose first. And choose he did—the very best room with floor to ceiling windows looking out over the lush stud on two sides.

A great treat for the children were the helicopter flights down to the stud. Although Robert and I had both gained our pilot licences at about the same time—he in New Zealand, I in Northern Rhodesia—neither of us had the time to put in the flying hours necessary to keep our licences current. Thus on these trips down to the stud, there had to be a pilot just as there had to be Paul's golden Labrador Charlie, probably one of the most experienced fliers in the canine world.

These were generally regarded as the years of Robert's greatest successes. And compared to his 727 jet aircraft—which as well as being fully staffed contained both an office and a bed—the helicopter was comparative small fry. His new home on the stud became one of several he owned: the Peppermint Grove house, an apartment in Melbourne where he and Janet would stay when visiting the children at school, and a new acquisition—the magnificent Grove House on four-and-half acres of prime real estate alongside Regent's Park in London. He also owned Trelawney Stud in New Zealand, Wallan Stud in Victoria, the glorious tropical Double Island off the coast of Queensland, and the prestigious Vasse Felix winery in Margaret River, Western Australia. Up in the north of Western Australia and in Queensland, twenty-three cattle stations covering over 1% of the total land area of Australia were gradually acquired under the banner of Heytesbury Holdings.

In the marketplace, he was considering his move on Australia's largest company BHP. Closer to home, he and Janet had amassed an amazing collection of art which they had started in a relatively humble way by buying 'what we like' back in the early days of their marriage in Darlington. I think they probably kept that philosophy right through—buying what they liked—but that didn't stop the collection growing at an enormous pace and becoming worth a considerable sum. A huge shed was built on the stud, designed to house both Robert's vintage car collection and to store such artworks that were not on display. Now they are exhibited in a gallery in East Perth where Heytesbury has its head office.

Two years after the house was completed, and with the words, 'While it may not affect the quality of the progeny, it's going to make the living for both horses and people a lot more pleasant', Robert made a decision to seal the gravel roads, something for which I will ever bless him. Replacing the original dirt tracks with eleven kilometres of road was a massive project that in later years the stud would not have been able to afford. The improvement in the appearance of the stud was almost incidental compared to the impact it had on laying the dust. But it looked good, too, since instead of grey bitumen, the new roads were coloured as close as possible to the original brown-red gravel. Lined as they were by Janet's tall and stately citidoras, which had long ago reached their full height, Heytesbury now had kilometres of leafy avenues.

We saw less of the children at this time, as they were now all in their teens and away at Geelong Grammar boarding school in Victoria. Although I was not to find out until years later, Peter and Simon particularly, perhaps because of their father's wealth, were subjected to a great deal of bullying at school. Unlike most children, Robert had loved his own boarding school days— despite the usual inedible meals and cold plunges in mid-winter—but his sons reverted to the norm. When they complained, Robert would give his standard answer: that it was 'character building'. But I wonder whether he knew the full story. In comparison with his own boarding school experiences, Geelong was not a happy environment for the older boys. It is testament to the character of the children that they stood up to it at all, and that they were able to put it behind them and go on to lead such successful lives of their own.

Catherine, too, had been schooled at Geelong, although her experiences were very different from those of the boys. One of her darkest memories of those years was not being allowed to leave school to watch Black Knight win the Cup. 'Instead', she said in an interview with *The West Australian*, 'I was sitting in a Japanese class listening to the radio.'

But the excitement of watching her filly Denver Dame win a

Entrance to the main house, Robert's 'big weekender'.

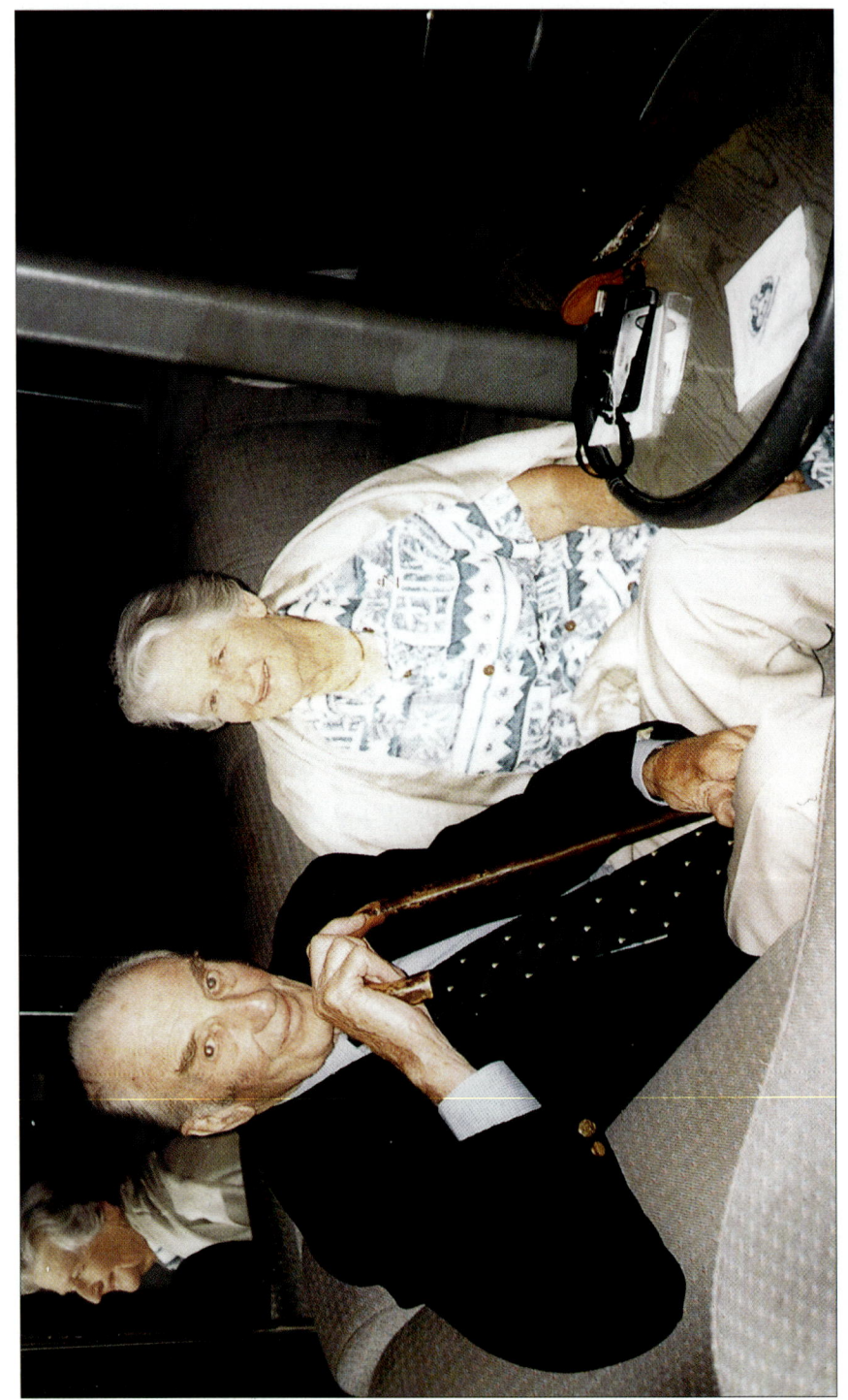

With Ronnie. Cruising on the *Sea Goddess* from Hong kong to the Great Wall of China, 1996.

With my lovely Savannah at Tshukudu Lodge in southern Africa, 2001.

I still ride at 87 years of age. Jacque and I, 2002.

My Zhivago would never forgive me if he didn't appear in this book, too!

With Lara, my ridgeback, during our morning walk down the avenue of citadora trees.

Our daily exercise!

Good morning, Lara!

major race at Belmont Park, Perth, three years later probably made up for it. Since she was only seventeen at the time and the rules state that one must be eighteen years of age in order to race a horse as an owner, she got around the problem by leasing the filly back to her father.

As she said, 'It all began when I was 12 and decided to buy Colorado Red from my brother Peter. Peter wanted $1,500 for her and I had to ask my father whether I could borrow the money from him'.

'Eventually dad and I made a deal that I could borrow the money at an interest rate of seven and a half per cent. I paid it all back when I sold her first foal, Haulrado (sired by Haulpak), for $4,000.

'When Denver Dame came along she wasn't the best-looking foal I had seen but I liked her. I knew a schoolgirl wouldn't be allowed to race a horse so I asked dad if he wanted to lease her from me.' (From an interview with *The West Australian*, Sept 28, 1987).

Although Robert had doubts about the filly's dam Colorado Red, in this case Catherine's instincts had been right: the filly's win in the Burgess Queen Quality won her $20,000.

For some ten years now the incredibly brave Haulpak had been standing beside Silver Knight. In spite of his accident, he learned to cope with a game leg that remained as thick as that of a young elephant. Robert had never varied in his opinion that if a horse were in any danger of suffering, arrangements should be made to have it put to sleep. Was Haulpak in pain? I often searched my conscience on this and I think that although from time to time he was, he was treated with painkillers the moment it became apparent. At the same time, he had one of the strongest wills to live that I have ever observed in an animal. Watching him serving a mare one day, his devoted stallion handler, a man by the name of Joe Murphy, muttered that when resting his bad leg on the mare's back 'was the only time the poor bugger gets any relief'.

But although Haulpak was a veritable cripple, like many humans he overcame his disability to lead a normal life. He really was the most courageous horse and it is significant that his march to supremacy on the West Australian Sires' lists for the 1985-6 season began in Australia's official Year-of-the-Disabled. He became the most sought-after stallion in Western Australia.

One sign of a good stallion is the passing on of his qualities— like colour and markings—to his progeny, the stamping of his foals with his own features. In his youth Haulpak had been black and almost all his offspring were born black like their sire. And although his accident turned his coat a dull bay colour, over the years it did turn dark again. I like to think he also passed on his special stamina. A Haulpak colt was well-prized and those of his progeny that weren't destined for the track actually made excellent dressage horses. Several of those that retired from racing were kept as studs both over East and in Western Australia. So Dashing was a successful horse that was sent East; Star Pack continued his career in Western Australia. Haulpak's bloodline was destined to continue.

It was just as well. From the meagre $250 we had so tentatively charged for his service in the early days, his fees had increased to the point where he laid claim to the highest stud fee ever charged in Western Australia. That fee reached a remarkable $15,000—still a record in this State. Another amazing statistic on Haulpak is that 10 per cent of his runners to reach the track became stake-winners, a statistic that compares favourably with the best sires in the world. But on the stud, he wasn't just one of the world's top sires, he was our much-loved pet and we spoiled him as much as we were able.

Meanwhile Pago Pago had been retired for three years and ever since Robert had decided he deserved a special diet and asked me to supervise his feeding, it had become a part of our daily routine to boil up his barley and slice the carrots. Ronnie cut him fresh lucerne each day using a sickle which, incidentally, we had had quite a job tracking down. And at least once a day I would stop by his stable, pat him and talk to him, and give him his carrot treat.

But as it turned out, of our three stallions, it was Haulpak who died first. We found him lying in his paddock late one afternoon and called an immediate meeting. That evening, it was an emotional group that congregated in my cottage—Ronnie, the resident vet, the stallion handler and myself—to put through a call to Robert who was on business in London.

Robert was as upset as we all were, but he didn't waver, 'Well, I've always said that if he appears to be suffering at all, we will have to put him down'. At these words, although we had known in our hearts that this would be the outcome, we were all most distressed. In fact, Joe was so upset that he threw his arms around me and started to cry. Over the years, I have seen how much stallion handlers worship the stallions in their care and Joe was no exception.

Haulpak was only sixteen years of age when he was put down on Saturday night, 4 November, 1989. On the following Monday, it was reported in *The West Australian* that '... the remarkable Haulpak, one of the most courageous horses that Western Australia has known was humanely destroyed at Heytesbury Stud on Saturday night, ending an era in thoroughbred breeding in the State'.

The upsetting events of life are rarely served up one at a time. Only forty-eight hours after the great Haulpak was put to sleep, there came another sad day and another tremendous loss when Pago Pago was found lying beside the water trough in his paddock. He was thirty years of age and during his lifetime, he had made a name for himself all over the world: through his stud name in America, through his progeny who went on to win in Europe, and of course he was a legend on home turf in Australia.

I feel sure that Haulpak's death contributed to Pago's in that Pago Pago knew he had lost his friend. Animals have that sixth sense and time and again through my life, I have found that they knew of or 'anticipated' events long before I did myself.

We received a facsimile from the Registrar of Race Horses in Sydney which reads: 'Re Pago Pago. I wish to advise that the name Pago Pago is a protected name and will never be given to any other horse in Australia. Regards, Keith McKay.'

Two small headstones were engraved with the stallions' names and dates and placed in the shelter of a grove of trees.

It was 1989, eighteen years after the founding of the stud. Two great stallions, two of my greatest friends, were dead. The silver-grey of the eucalypt leaves shimmered in the breezes along the avenues. If we had but known it, the winds of change were stirring. Heytesbury was heading for another era.

CHAPTER THIRTEEN

Storm Clouds Bursting

Perhaps it is inevitable that years of plenty are followed by years of drought. Certainly it was so in the case of Heytesbury Stud…and of course my own life, being so closely entwined with that of the stud, rapidly headed downwards in the same convoluted spiral.

Robert died on Father's Day, September 2, 1990. The sense of shock and loss that hit me simultaneously on that day is something I shall never forget. He had suffered a massive heart attack up at the main house on the stud and was dead by the time he arrived at the hospital. The disappearance of Simon had affected me deeply; the death of Robert was just too unreal to grasp. My two sons—the sun and the moon of my life—both gone, and I, their mother, still living. It went way beyond the actualities of reality and became something that I could not understand.

But when you think things can get no worse, you find out they can. I was not the only one who had difficulty coming to terms with Robert's death. Without Robert, circumstances changed literally overnight. I don't think anyone could visualise the stud without him. No one could cope. The $800-million massive Heytesbury empire was threatened from the bedrock upwards. Robert had had no contingency plan in place. Perhaps my son thought himself immortal. I remember his saying at one shareholders' meeting, for which he had arrived late due to an accident, with a cynical smile, 'Reports of my death have been grossly exaggerated...' He left no will—'because things change too quickly for a will'—and no one properly trained to take his place. The company staggered. The stud staggered. We all staggered.

Bravely, Janet stepped in the following day and took over the running of Heytesbury Holdings. She did her best, but she was not Robert. The banks knew she was not Robert. He had a lot of borrowings; the banks started to call in their loans. The accountants' focus began and ended with the bottom line and that line was underscored in red iridescent ink. Head office swung into action. Sell, sell, sell...was the refrain. Immense confusion, the result.

The leader and planner that Robert was could not be duplicated. You can't run a stud from head office, but unfortunately they tried. Perhaps the most disruptive and destructive action taken, certainly initially, was the indiscriminate sacking of staff.

At that time we had a manager well-liked by the people who worked for him. He was fair and well-respected not only for the time and effort he put into the stud, but also for his knowledge of horses and his business acumen. Head office arrived at the stud not long after Robert's death, called him into the stud office and sacked him on the spot. No explanation. 'Pack your bags,' they said, but a little less politely. He still looked dazed when he came to find me. A moment later he was gone.

I think it was this action, which came with no reasonable explanation, that started the confusion. Inevitably the other staff reacted by panicking and the result was similar to the collapse of a house of cards, except that this was no house of cards. This was Heytesbury and the people who were leaving were dedicated workers with real lives that had made a solid contribution to the well-being of the stud.

Now the stud was regarded as a luxury. There were those who hoped that 'within the next five to seven years there won't be a bloody horse left on this property'. There were others who went out of their way to try to make that a fact.

At the time of the culling, we had 162 mares on the books between Heytesbury and Trelawney Studs. The sales were ruthless and indiscriminate. Some of our best broodmares were sold from under our feet. Among these were the ones Robert had brought in from all over the world in his ever-continuing efforts to improve bloodlines. There appeared to be no thought given to the fact that their progeny were winning on the tracks, and that they were—in fact—good money-spinners. Others were some of our favourite old mares. Horses were leaving the stud daily until eventually only about forty mares remained. There were fewer horses sporting the Heytesbury maroon and white colours on the track. The horses racing overseas were sold. 'All too costly, much too costly,' was the reply when we begged for some explanation. We lost clients. The anguish I felt seeing the destruction of an enterprise to which I had made quite a sizeable contribution over the years, was deep and hurtful. I don't know what I would have done without Ronnie's quiet strength.

They got rid of land, too. Some of the acreage which had provided some of the best pasture was sold. Little Heytesbury across the road, sold. Over East, the picturesque stud with the Cape Dutch style buildings at Wallen in Victoria, sold. Trelawney Stud, birthplace of our first stallion Silver Knight, eventually sold. Westover down at Benger, less than an hour from Heytesbury, where the mares and their foals were spelled so successfully over the summer and autumn, sold. The list went on and on. All sold. The stud was falling apart in front of our eyes.

With the sales came memories. Westover had been named by Robert's cousin, Bridget Holmes à Court. Robert brought her out from England for a holiday shortly after he had purchased the Benger property and when she saw the mix of hilly country and flat irrigated land so ideal for horses, it reminded her of a farm in England: a portion of the original Heytesbury estate called Westover. So the new acreage was christened Westover. Bridget was a real character. When Ronnie and I were in England on one occasion we went to see one of our horses—Girl Friday—run in a rather important race. We were craning our necks to get a good view, but this was not good enough for Bridget. Off she went through the crowd, using her walking stick to good advantage when people didn't move out of the way fast enough. I can still hear her *don't-you-argue-with-me-young-man*! They didn't. The filly ran a very good second and as it happened we had backed her because Ronnie thought it might spur her on. We bought two paintings that still hang on the walls of my cottage as a result of that win. And it was thanks to Bridget that we saw it at all. Memories of a happier time.

A friend came to visit, 'What on earth's the matter Ethnée? When we turned in at the gates we could feel the atmosphere? Something awful's going on. What's happening?'

The turnover of staff had become ridiculous; the tension an almost tangible part of the daily atmosphere. No one was staying. It was all most unsettling. I had been here from the very first days of the stud and it was a part of me.

I stand a little shakily in the centre of my kitchen and my gaze falls across a sack of carrots. I decide to give the horses their treat a little early. Something steadying. Normal. I chop up a handful, grate others for Nolasque who is getting on in years now and has very few teeth left, tear pieces of crust for the little teaser stallion who loves both bread and carrots. Place the treats in a plastic bag.

We leave Ronnie writing and dozing by turns on the back porch. I jump onto my tricycle, and with my ridgeback trotting beside me, go off to do my rounds. When Nolasque sees us coming, she nickers softly and comes to greet us. She is just a bit grumpy today and snorts at the grated carrot so that it spills and she has to search for it on the ground. Animals, particularly horses, are the first to pick up changes in atmosphere. They know I am upset just as I know they are not quite themselves.

Next we go up to the paddock beside the serving barn to give the bread to the stocky little Welsh Mountain pony that we use as a teaser during the stud season.

I complete my tasks, but I am not at all sure I am feeling better. I race back to Ronnie and find him still sitting under the pergola. He is so calm. The peaceful centre of this senseless cyclone. He holds me. Takes me by my two hands. Leads me back to the porch. Checks the sinking sun for permission to pour us each a drink and puts the glass in my hand. He makes sure that my ridgeback is with us. And then he picks up one of his favourite pieces of prose that is lying on a nearby table. My heart is breaking. My heart is melting. He starts to read "Youth".

> *Youth is not a time of life; it is a state of mind. It is not a matter of ripe cheeks, red lips and supple knees; it is a temper of the will, a quality of the imagination, a vigour of the emotions. It is a freshness of the deep springs of life.*
>
> *Youth means a temperamental predominance of courage over timidity, of the appetite for adventure over the love of ease. This often exists in a man of fifty more than in a boy of twenty.*
>
> *Nobody grows old by merely living a number of years. People grow old only by deserting their ideals.*
>
> *Years wrinkle the skin; but to give up enthusiasm wrinkles the soul.*

Worry, doubt, self-distrust, fear and despair—those are the long, long years that bow the heart and turn the greening spirit back to dust.

Whether sixty or sixteen, there is in every human being's heart the lure of wonder, the sweet amazement at the stars and at star-like things and thoughts; the undaunted challenge of events, the unfailing, childlike appetite for what next, and the joy of the game of living. You are as young as your faith, as old as your doubt; as young as your self-confidence, as old as your fear; as young as your hope, as old as your despair.

In the central place of your heart is an evergreen tree; its name is Love. So long as it flourishes you are young. When it dies you are old. In the central place of your heart is a wireless station. So long as it receives messages of beauty, hope, cheer, grandeur, courage and power from God and from your fellow men, so long are you young.

We are both quiet when he finishes reading. My proud peacock and his two faithful guinea fowl walk jerkily past. The phrase 'the game of living' reverberates in my mind. A little of the distress starts to dissolve.

Once the reorganisation was complete, there was very little left of the old empire.

The truth of the matter was that the stud had been Robert's love and hobby. When he died he had four studs, three in Australia and one overseas. He had dominated the West Australian Breeder of the Year stakes for years past. For nine years Heytesbury had been a winner of the 'Sire of the Year' award.

Three times the stud had been awarded 'Broodmare of the Year'; small wonder that the progeny of Heytesbury broodmares were winning on tracks all over the world, in countries like Singapore, Malaysia, Hong Kong, Zimbabwe, South Africa and the United States. On their home turf, the stud has produced winners of some of the country's most prestigious races, among them the Group One races like the WATC Karakatta Plate, the WATC Derby, AJC Epsom Handicup, the VRC Australia Cup as well as scores of winners of other Group and listed races.

He had lavished millions of dollars on improving the bloodlines that led to these accolades, never hesitating to pour money into the stud to create the best thoroughbred stud in the southern hemisphere. His credit was such that all he had to do was to front up to one of the several banks he dealt with and say, 'I want to buy another stallion. I need a million dollars'. Now that he was gone, spending on such a scale was not only impossible to keep up, but very difficult to justify. Meanwhile Janet was overloaded with the magnitude of the sudden responsibility, and the children were too young. There was none of the family input so critical to a venture of this nature.

This is how the ambiguous state of affairs that Robert left behind affected the stud and our lives within it. But the parlous state of the whole Heytesbury empire made our catastrophe seem relatively small.

It cannot have been easy for Janet. Despite the multitude of problems she managed to resolve, there was still a minefield of ambiguities and uncertainties. Robert had left an empire which employed over 2,500 people all over the world and which, although asset-rich, was based on debt. And he had prepared no one to take his place. Although it had been assumed that Peter was being 'groomed for the top job', and that as eldest son he would naturally follow in his father's footsteps, this was not to eventuate. Partly I think this was because of the sudden circumstances of his father's death and the fact that he was only twenty-one years of age at the time. Also I think Peter felt he needed to complete his studies and fulfill his theatrical ambitions before assuming a corporate role. Certainly he had attended company meetings and acted as a stand-in for his father while he was still a university student. But in the confusion of the aftermath of Robert's death, it was Janet who took over control of Heytesbury Holdings.

As it turned out, Peter decided to return to studying his law degree at Oxford and appeared to be very happy when Ronnie and I visited him there a short while later. He had elected to stay in 'digs', which were not your normal idea of digs at all, but a rather large and comfortable house. It was separated from the university by an open space of lawns and huge chestnut trees, with an almost rural feeling brought about by the cattle grazing in an adjoining paddock.

In fact, all the children worked hard at their studies in those early years after their father's death. Simon had started an arts degree preparatory to undertaking law at the University of Western Australia and Paul was in his final year at Geelong Grammar. Catherine, who had graduated with an arts degree from UWA, had decided to spend the following two years doing postgraduate studies at Oxford.

Meanwhile Catherine's 21st birthday, not long after her father's death, produced the following letter from Ronnie:

My dear Catherine,

Ethnée and I give you a small present for your 21st birthday and, with it, our very best love.

A 21st birthday is an important occasion in anyone's life and I have therefore decided to write you a few lines on my thoughts about you and life in general.

Since I first saw you as a kid in the swimming pool some twelve years ago, I have had a great affection and admiration for you. I have seen you competing in the Show Ring with grace and elegance and watched you accepting awards on Race Awards night with poise and elegance. Thank goodness I've never seen you in the hockey field. Your father always said it was a frightening experience!

May I say a few words about 'life in general'? I think three things are important: Know thyself, know where you are going, and have integrity. The last is very important because, with the shield of integrity, however the fates may play, you march always in the ranks of honour.

In your journey through life I am reminded of the words of a poet:

> *We are the Pilgrims, master; we shall go*
> *Always a little further, it may be*
> *Beyond that last blue mountain barred with*
> *   snow*
> *Across that angry or that glimmering sea—*

I am sure that you will ride out any storms, and have many happy voyages through that glimmering sea.

Bon voyage, Catherine
Ronnie

The storm had broken. The first empire had come and gone in the way that empires do. Now it was time for us all to voyage through that sometimes angry, sometimes glimmering, sea.

CHAPTER FOURTEEN

Pots of Gold

The storm passed and the sun shone again...in short rather watery bursts at first, but as the years unfolded, things continued to improve.

It was Ronnie's suggestion that we spend less time at the stud and instead indulge our enjoyment of travelling and visiting our many friends around the world. I think this was his way of forcing me to relax. He knew very well that while I was around my animals I would be busy not only from dawn to dusk but very often through the night as well, particularly during the foaling-down season. The atmosphere at the stud was still thick with uncertainty and getting right away from it all was one way of allowing the balance to right itself with time.

Consequently the next ten years were golden indeed. Apart from the unceasing delight we found in each other's company, we looked outside our normal lives for different things to do. We bred racehorses of our own which we raced in Ronnie's regimental colours of green and blue with a silver stripe. We went back to Africa several times. We travelled the world from the Antarctic to Alaska. Our beloved pets Zhivago and Lara came to live with us. I wrote my first book and travelled around the globe promoting it for a furious twenty days.

But this period produced something even more valuable. It's an ill wind that blows nobody any good, or so the saying goes, and out of all the trauma that still came roaring down over the ranges to meet us from time to time, came something rather wonderful. Although I had lost my two sons, I still had my grandchildren. They had never stopped coming down to the stud for their holidays and Ronnie had always been a grandfather. In fact, so well had he fulfilled that role that the children actually came up to me and thanked me for giving them a grandfather. So although I would always be busy, because I'm not sure I knew how to be anything else, the after-shock of all we had been through allowed me to slow down just long enough to start appreciating what I had instead of missing what I had lost, and to spend more time with the children when they were at the stud.

They were, of course, children no longer, but young adults all doing well for themselves in the very different careers that matched the differences in their characters. Over the years that had passed since she received her 21st birthday letter from Ronnie, Catherine certainly demonstrated that she had what it took to 'go further' in a number of different directions. 'They don't give you much time to rest at Oxford'—she wrote. Her study programme alone could not have been easy—a master of philosophy majoring in economics and politics—but this was only part of her achievements in the Nineties. While she was at Oxford, she was not only awarded a 'Blue' for rowing, but also became interested in competing in a unique field—that of the pentathlon. Before long this interest turned to enthusiasm and then passion, and almost before we knew it, she had become a pentathlete. We had followed her successes at gymkhanas and pony club events so often during her teen years that this provided us with an added interest.

One thing is for certain: you don't get to be a top pentathlete by being a wimp. The five disciplines of the pentathlon are extremely demanding and diverse. Shooting, swimming, fencing, jumping an unfamiliar horse, and running events all have to be completed within a 12-hour framework. The demand is not only for expertise in each sport, but a high level of stamina and endurance is critical. Most people arrive at pentathlon level from a background of proficiency in at least one of the sports. In Catherine's case, she protested that she was 'equally average' at all five, but that her show-jumping background had probably helped her more than anything else. Having three brothers and a huge acreage on which to run and swim probably helped a great deal too. Whatever the reason, she was hugely successful and over the years won many accolades within this difficult competition.

Regretfully our trips didn't happen to coincide with her sporting appearances, but we were able to share in her progress through her letters and newspaper reports. At the beginning of 1992, she wrote, 'I have been training quite hard for the Modern Pentathlon—I hope to make the varsity team, it would really mean a lot to me'. She not only made the university team, but later that year she qualified herself to enter the national championships and just over a year later was competing in the world championships. In 1993, she came sixteenth in the world and wrote back 'obviously I was ecstatic'.

We were ecstatic, too. She loved the pentathlon because it was competitive and she had the necessary spirit for competition, but she also loved meeting people from all over the world. Her Oxford studies completed, she was not long back in Australia before she was off again on the international circuit: visiting New York for the opening of Peter's play, and then heading off for London, Paris, Rome, Hungary, Poland, Vancouver, Greece and then back to England for the world championships in August. She maintained that 'living out of a suitcase for four and a half months will probably be the biggest challenge of all! As Dad would say—it'll be "character-building"!'

By this time, Peter had graduated from Middlebury College in Vermont in the United States with a degree in economics; his years at Oxford had yielded a law degree. He had also decided that his initial foray into investment banking in New York was not for him but that a career within the theatrical world was. And so he entered the theatre.

The first time Ronnie and I saw him on the stage was at the Edinburgh Festival Fringe in 1992 where he gave an impressive performance. We went the next morning to see him in another show, *Breakfast with Shakespeare*, a most amusing play held in a little playhouse in the centre of Edinburgh not far from a club where Ronnie and I were staying, and which served excellent coffee and croissants during the production. Somewhat to our surprise, over dinner that night, Peter persuaded us not to come to that evening's performance. 'Don't worry about coming tonight', he said. 'You'll find that it's only more of the same...' He possessed his father's power of persuasion and so we had an early night that we didn't really need.

But actually it was rather *less* than '*more* of the same'. When we opened the newspaper the following morning there was an article devoted to Peter's performance...on stage and nude...wearing only a pot plant. I suppose he had thought he was sparing the sensitivities of the older generation by suggesting our non-attendance.

Peter's own theatre company, Back Row Productions, debuted in 1995, and I guess the 'pot-plant' performance had foreshadowed that the market he would aim for would be young and adventurous. It had taken a little time before he became successful, but once he was, there was no stopping him. Although I suspect that even if he had not made a success of the theatre, he would still have been happier than working for somebody else. Like his father, he very much preferred working for himself.

While Peter described his venture as 'a risk-taking business' as opposed to the established theatres like Stoll Moss owned by his mother, it was large enough to warrant offices in the major cities of New York, London and Sydney.

At the 1995 Edinburgh Festival Fringe he produced a programme of five first-class shows, many of which received excellent reviews from national critics, but none more so than *Tap Dogs*, the undisputed hit of the Fringe that year. *Tap Dogs*—produced in cooperation with the Sydney Theatre Company—was tap transformed, a non-stop dance tapped out with steel-capped work boots on a mainly metallic construction set. The dynamic all-male group literally kicked its way to become the hit of the season with this production. A sell-out in Edinburgh, the show toured all over the world including Germany, Japan and Canada, with the fierce magnetic energy of the players attracting rave reviews wherever they went.

Instead of going to Middlebury College like his brothers, Simon meanwhile had decided to go to the well-known and respected Dartmouth—an Ivy League college situated on the New Hampshire/Vermont border in the United States. The university is set in a most incredibly beautiful part of the world which attracts 'leaf peepers' from all over the country as the trees change colour in fall.

He had met Katrina, his wife-to-be, while they were at school in Geelong…and for six years they had been together whenever possible while pursuing diverse careers. Simon's studies and work had taken him across three continents: Australia, Asia and the States. Katrina's musical background resulted in her living for a year in that great music capital, Vienna, in order to study with one of the city's leading singing teachers before she also decided to go to the States to continue her studies.

'But at first," she says, "I managed a restaurant for a year in 42nd Street, New York…a very trendy place to which all the celebrities flocked. I used to spend eight hours on the bus to get to Simon in New England…sometimes it was late at night and I would be the only person left and so it would get a bit scary…'

It wasn't long however before she again auditioned with a famous singing teacher, was accepted and took lessons. Although neither Simon nor Katrina may have known it at the time, they were on the brink of one of the most challenging and professionally exciting periods of their individual careers.

Meanwhile Ronnie and I were travelling the world to some of the most fascinating places imaginable, and almost all our holidays were animal-oriented in one form or another. We went on fishing holidays to countries as far away as Alaska—where I also managed to catch a sixty-pound salmon which impressed all those fishermen no end!—and to Ronnie's old haunts in the Outer Hebrides. One of my fondest memories of Alaska was sailing past two polar bears playing on the far bank of the river, their antics captured by the clever brushstrokes of artist Shirley Cant. Ronnie later wrote to this artist and bought the painting as a present for me. It hangs on the wall of my cottage and remains one of my greatest treasures.

We spent three wonderful weeks on an expedition to the Antarctic cruising among dozens of whales and spotting hordes of rather supercilious penguins. But although the wildlife of Antarctica was so different, it was the sheer beauty of the icebergs that dominated this trip. Standing with our arms around each other on the starboard side of the ship one evening, we passed close to one of these bergs just as it caught the full reflection of the dying sun. The brilliant orange-pink against the charcoal of the sea and sky remains vivid in my mind.

Africa, of course, drew us back time and again. We visited Kenya and revisited Botswana and Zambia, taking trips along both our mutual and individual memory lanes and deriving enormous pleasure from viewing and visiting the habitats of our beloved wildlife once again. One day we drove out to Ronnie's old home at Blue Lagoon. Ronnie had decided the time had come to clear out the pictures and furnishings. We were a little worried as we didn't know quite what to expect of the property, particularly as the road up to the house had deteriorated badly. But actually, and somewhat to our surprise, very little had changed. Ronnie's old house boy met him at the front steps and the contents of the house were in quite good order. Another few nostalgic days were spent at Chobe, which had been my home from 1959 to 1964, and which had since been turned into a National Park.

Very different were our trips to the United States and perhaps of all we saw of that great continent it was Kentucky, and the numerous horse studs we visited in that state, that provided the greatest attraction because no matter how far we wandered, Heytesbury was always in our thoughts.

Back home our animals continued to enrich our lives. Probably one of the best presents I have ever received was Zhivago, the Russian Blue cat Ronnie gave me for Christmas in 1992. Anna-Lou—who has worked for me now for over twenty years—had a lovely mother cat called Blue that produced five kittens. They were all splendid lively little things, but one was more adventurous than the others. This was the little fellow Ronnie picked for me. He started off his life with us in a little basket with a blue rug, but needless to say, this didn't last longer than a couple of nights; he found my bed and electric blanket much more to his liking.

At that time we still had my ridgeback Maggie May—named after the yacht in which Simon travelled around the world—and cat and dog made friends immediately, accompanying us on our walks around the stud and curling up together in front of the gas fire on cold evenings. We found before long though that the adventuresome temperament Ronnie had spotted from the beginning was to lead Zhivago into some awful fixes. Sometimes he would climb a tree to find he had gone too far and was unable to turn and safely come down. Another time he went rushing into a rather long drainpipe. We couldn't see him from either end and it took a great deal of calling and poking a stick down one end before he emerged from the other covered in leaves and dust.

One other adventure was almost his last. Although he was normally home before dark, on this particular night he couldn't be found anywhere. It was not until the next morning that Anna-Lou happened to find him lying in a flowerbed just off a busy road not far from the cottage. From the unfocussed stare of his eyes and the difficulty he had breathing, he was obviously badly injured and in a state of shock. As carefully and as quickly as possible we carried him to the car and raced down to the vet in Pinjarra. Several x-rays later, we were told that he had internal damage and a broken jaw. He was treated for shock and pain and put into a warm bed by the heater. The prognosis, we were told, was not good. By this time, I had known Zhivago for two years. I loved that cat. We both did. We were not prepared to accept that he was going to die.

We consulted an orthopedic veterinary surgeon in Perth who again advised rest and treatment for shock. But he didn't look very hopeful either and I simply was not prepared to give up. We decided to have a third try. We took him to another clinic in the city for further x-rays and then I almost wished I hadn't when we were told the extent of the damage and of the necessity to operate. I don't like operations, but sometimes there is little choice. I had a feeling we didn't have much time to waste. The chance of Zhivago fully utilising his nine lives was looking less likely by the minute. We decided to take a chance and opted for the intricate, and consequently extremely expensive, operation. Thank goodness we did because it was a complete success. After a few days we were able to take him home, and six weeks later he had fully recovered.

Did this rather severe lesson teach Zhivago to be less intrepid? Perhaps. Although on the surface he was the same mad cat he had always been, galloping with the dogs and racing up the tallest trees, sometimes in the way he looked at me I had the feeling that he remembered a little of the pain we both felt and tempered his leaps with just a smidgin of caution.

❖

In 1995 we were back in London again for Ronnie's 90th birthday party. Of all the places it could have been held, there was nowhere more appropriate than the gracious ambience of the Cavalry Club at 127 Piccadilly, right in the heart of the city. For someone with Ronnie's background the guest list was appropriately eclectic. All Ronnie's friends who helped celebrate his 80th birthday at a shooting party in Scotland were with us again, this time to partake of a sumptious lunch served against a backdrop of glorious oil paintings and crystal chandeliers. There were fishing and polo-playing friends from Scotland and England, friends from army campaigns fought in Burma and Ethiopia, friends from Zambia and from Western Australia, as well as members of our own families. Ronnie's step-son from his marriage with Erica, Michael, and his wife were there. Catherine, too, was able to attend. The list included a number of distinguished guests. One thank you letter said it all rather succinctly: 'We were truly a very elegant group!'

Key among the telegrams read out was one from Princess Diana—Colonel-in-Chief of Ronnie's regiment, the 13th/18th Royal Hussars:

> *Many congratulations. I send warm good wishes on your 90th birthday and hope that all assembled at the Cavalry and Guards Club make sure that this special occasion is celebrated in traditional style.*
>
> *Diana*
> *Colonel-in-Chief*
> *The Light Dragoons*

It certainly was.

❖

It was not until 1997, well after the death of Maggie May, that we decided Zhivago needed another ridgeback companion. Neither Ronnie nor myself felt like puppy-training, so we decided we would look for an older dog. Eventually we found the pet we were looking for. She was a year-old wheaten-coloured female with huge brown eyes and the overall conformation one would expect from the daughter of a champion. Rhodesian ridgeback breeder, Robyn Black, brought her to us one Sunday afternoon. She had been named Goldie, but we didn't much like it. Since we already had Zhivago, to Ronnie she was a Lara—so Lara she became.

Although we fell in love with her immediately, ridgebacks are thinking dogs. They very rarely obey an instruction unless they have thought through the consequences of treats that might be given or withheld as a result, and Lara was no exception. Although she made it clear she had no wish to follow Robyn back to Clackline, each evening for the first few nights she was with us, she sprang onto the chair in which Robyn had sat, stared solemnly at us with that long serious look she has and then curled up with her nose snuggled into her paws and went to sleep.

But once she decided to give her trust, the love on both our parts became unconditional and infinite. We would do anything for each other.

I am so close to my animals that I can read their minds and I know they can read mine. Recently I was at my friend Alison's house on a visit to Perth. Alison just happens to be Lara's closest friend, too, probably because of her generosity with the contents of the ginger biscuit tin. This particular day, I was just thinking it must be about time for my car to arrive when Lara sat up and looked out of the window and all but said to me: 'Here's the car. Come on!' It must have been less than a minute later that Jacque—who has been at the stud for many years and whose principal role is that of night-foaling supervisor, but who now often drives for me—turned the car into the driveway.

Dogs, cats and horses don't need words to communicate and they don't mess up the business of communicating quite so thoroughly as humans. Rarely do I have to order my animals to do anything. I've found that once you have gained the respect of a pet, all you have to do is to ask, not command, and more often than not they will readily obey. They also have such individual personalities. Lara, for instance, knows most of my friends by name and shows enthusiasm for a visit commensurate with the quality of the biscuits they provide. Zhivago gets most upset when it's vacuum cleaning day and asks to be let outside. It works in the reverse order too: If I'm upset, they are instantly at my side.

I find great satisfaction in the companionship of my pets. In short, I adore them. When they die, I suffer tremendous loss because I have lost a friend. I miss them deeply and grieve their departure for a long time. Some folk see them as 'just a dog' or 'just a cat'.

> *And when at last there comes the dreaded break,*
> *Someone who never owned a dog or cat*
> *Will smile and say, "Rather a fuss to make*
> *And not a very handsome dog at that".*

That type of person I do not understand.

The golden years produced a golden shower of weddings. My three older grandchildren and their partners expressed their individuality in three beautiful ceremonies spaced two years apart.

Although they were both still living in New York, Peter and his American fiancée Divonne Jarecki decided on a bush wedding at Glenprairie, a Heytesbury cattle station in Queensland. Glenprairie is about two hours' drive from Rockhampton and the Heytesbury pilot was kept busy flying literally loads of people, flowers, champagne and wedding presents from one place to another. The pre-nuptial celebrations lasted for three days while everyone got to know each other through a number of organised and more casual events such as looking around the station, horse-riding and, on the eve of the wedding, at a barbecue held at sunset on the banks of a nearby river. On June 20, 1995, Divonne and Peter were married in an unusually beautiful ceremony held under a large tree in front of sixty guests from all over the world.

When Simon and Katrina von Möller-Harteneck decided to marry in 1997, their choice of setting was equally unconventional:

> *You might ask why we chose to have our wedding in Greece. The answer is simple: when you invite people from places as far flung as Australia, Norway, Austria, France, the Netherlands, the Czech Republic, Italy, Switzerland, Malaysia, Thailand, Liberia, Hong Kong, Taipei, the United Kingdom, the United States and Canada, nowhere is central. Unavoidably, most people have to travel, so we chose a special place we are sure you will love.*

That 'special place' was the island of Hydra — once again a logistical challenge in terms of organisation. Again the wedding celebrations took place over several days with the ceremony itself held on Friday, June 20, tucked between a selection of festivities and sightseeing and one of the island's biggest local festivals.

Catherine and Robert Mather were married at the stud itself, and the ceremony was held on the slopes of the ranges looking down over Heytesbury. During the pause in events that occurs at the signing of the register, soloist Sara Macliver kept us entertained with her wonderful voice; at one point I wondered whether the sound of Mozart's *Alleluiah* was going to lift the eagles' eyrie off its perch. Afterwards we came down to the main house where a huge air-conditioned marquee had been set up on the lawns to provide dinner and dancing for as long we could keep going. And that was a fair time. It was well past 2 o'clock in the morning when Paul came up and swept me onto the dance floor. The sounds of merry-making were to merge with the early dawn chorus of the Heytesbury birds.

Despite my fears the Nineties had produced its share of rainbows, and there was a pot of gold at the base of almost every one.

CHAPTER FIFTEEN

'I Shall Not See the Shadows...'

My change of focus from the stud to our personal lives for much of the Nineties is as deliberate in these pages as it was in fact. I thank God for Ronnie. If Robert and Simon were my sun and my moon, surely Ronnie was the axis on which my world spun? He seemed to know instinctively when to suggest another trip, and we travelled the world from north to south, east to west. He was the calm at the centre of every cyclone that had shaken the stud to the roots of its majestic eucalypts. And over the years he had even found creative ways around my lack of cooking acumen, discovering places that would put together the most delicious take-away meals long before takeaways became a way of life.

It would not have been possible for me to bear a closer involvement with the changes that had been consistently taking place at Heytesbury Stud as it continued to lose the honourable place it had so painstakingly etched for itself in the competitive world of thoroughbred racing and breeding. Ronnie knew that. But beyond the obvious, and unbeknown to me, change was again stirring. If I had been able to gaze into the future I would have seen the signs that the stud was easing from the first to the second empire.

It started innocently enough one day in 1998 when Ronnie

suggested to Paul that they go into partnership by buying a filly together—one good enough to race. 'And not only good enough to race,' he said, 'but, more importantly, to *win* races.' 'To buy this filly,' he added, 'I am happy to pay up to $150,000.' So Paul sought the advice of William Hastings-Bass—now Lord Huntingdon, the English trainer whose clients had included the Queen and Robert—and together they went to the Eastern States to find a filly.

But Ronnie wouldn't have been Ronnie had he not added advice of his own. 'Mind that you choose only the best. Bear in mind that she should have a fine pretty head like a housemaid, a big strong bottom like a cook, and she must walk like a hooker...'

With that advice colouring his cheeks, Paul went off with the two great trainers—George Hanlon and William—to look at over a hundred fillies, and out of them all, to spend the sum of $125,000 on Johann's Dream, a beautiful bay filly with an enormous blaze. But Johann's Dream not only possessed the face, body and ease of movement to fit Ronnie's physical description, she turned out to have another essential quality, that secret ingredient which the best conformation in the world cannot make up for: the will to win.

On the day of the arrival of Johann's Dream, Ronnie turned to me, 'My darling Ethnée. They've done well. She's very lovely and I'm well pleased with their choice. Now I am going to give you my share so that you will be in partnership with Paul.'

It was this present, this interest that Paul and I were to share so fiercely while Ronnie looked fondly on, that bonded my youngest grandson and I so closely and which I believe instigated a large part of Paul's later interest in the stud.

Paul came more and more often to the stud as the century drew near its end and the friendship he had formed with Ronnie as a boy grew even stronger. Ronnie was nearly ninety-four years of age and although most days he still managed his twice-daily walk, his decreasing strength curtailed the activities they had so enjoyed together over twenty years. Now instead of skinning rabbits or fly-fishing the lawn or dam, they talked. They spent hours together chatting over the state of the stud and what could be done: Ronnie gently raising a point and Paul's enthusiasm and creative mind rising to suggest a solution.

At about this time and to my great delight, my first book, *Undaunted*, was published by Pan MacMillan. The promotion of the book meant I was whisked from Perth to the east of Australia for an incredibly speedy tour of three cities in three days, with a perpetual and merciless round of interviews and lunches. I had barely arrived in Sydney before I was whirled from one set of offices to another: to speak on radio, appear on television, talk in public.

My first attempt at public speaking was on a podium at the Sydney Hilton, at a lunch for 500 people. I quaked with nerves I'd never experienced before. Bravely note-less, I heard Robert's words at the edge of my already foggy mind: 'You should never, never, read a speech'. That was all very well, but I couldn't help wondering whether Robert had ever had an attack of nerves like this. In that situation, all you can do is go ahead, which is what I did. I eventually relaxed into the situation so much that the next thing I became aware of was my guide from Pan in the front row making signs to indicate I'd said enough. The rest was wonderful. We were put up in the best hotels and treated royally.

When we arrived back in Perth, our hotel happened to be the Parmelia. Because Ronnie had used the hotel as his club for so many years, I was given the best and most enormous suite. In between my various speaking engagements was the launch at the Parmelia, which proved so successful that one of the staff had to race out to Dymocks to bring in additional books for signing.

Now the time arrived to launch the book in Africa. I was looking forward to the trip, to meeting up with old friends. But not to leaving Ronnie.

'It will only be for a week,' I said to him, actually trying to reassure myself.

But as he had done all the way along, right from the moment he had set his sights on where and with whom he would like to spend the rest of his life, Ronnie surprised me.

'I don't want you to rush back. You need the break. I'd like to give you a first-class around-the-world ticket. Since you will be in Africa, you might as well continue your journey around the globe and visit all the children on the way.'

So I did. I went on to London to visit Catherine and their little girl Emma, on to New York to stay with Peter, Divonne and their very young twin boys. On one hilarious occasion, my sister-in-law Josephine and niece Caroline came to New York from Washington D.C. for dinner and we laughed the evening away. I went down to San Francisco to stay with Simon and Katrina and spent another couple of days with another niece in Los Angeles. Ronnie's present of a first-class ticket meant that during the tedious fifteen-hour flight back to Sydney, I had the most comfortable bed on the aircraft. And then back to Perth and home.

Around the world in twenty furious days. The pace was fast, even for me.

I needed to go, but I was just as glad—even more glad—to be back home. How Ronnie had aged in that short time. Or was it only that being away from him had made me realise how frail he was becoming?

Ronnie had once said that one of his regrets about death would be that he would no longer be able to enjoy the delights of the world of nature with which he had allied himself so closely in life. I heard the catch in his voice and I knew that he knew the time was fast approaching when he would leave this world when he quoted a line from one of his favourite poems, 'I shall not see the shadows, nor feel the rain...'

It is the night of August 26, 1999. Ronnie and I have had an evening together of food and wine, lots of chatter and laughter. Ronnie stops. He looks a little confused for a moment. Then reaching for my hand, he says quietly, 'I'm ready to slip away...' I hold his hand tightly. Lara whines and puts her head on my knee. Zhivago jumps onto my lap. The little circle is close and complete. Ronnie dies in the early hours of the 27th just as he has lived his life. With love. With integrity. And with the greatest dignity.

The void he left in my life and in that of the animals was immeasurable. We missed him sorely: his love, his support, his presence, his unfailing sense of humour. His *being*. I have a feeling that both Lara and Zhivago knew with more certainty than I that he had gone to a life in another world. It was a long time before they would even leave my side. Eventually Zhivago regained some of his independence and would leave me for short periods, stalking the house in a *dwaal*, as we would say in Afrikaans, refusing to go anywhere near Ronnie's room. It was not until a full two years later as I was passing Ronnie's den that I noticed a Russian Blue tightly curled in what used to be his favourite spot on Ronnie's chair.

Condolences came from everywhere. Ronnie's nephew Benedict dealt most efficiently with the numerous requests for information for obituaries, which were published in some of the leading UK newspapers like *The Times* and the *Telegraph* as befitted the war hero he was. The *Annandale Observer* devoted a special news feature to 'the last laird of one of the district's stately homes', at the same time paying tribute to Ronnie's international reputation as a conservationist and lover of wildlife. At a cremation ceremony held in Perth, one good friend said sadly: 'The mould is now broken…we will never see his like again'.

Robert and Ronnie—two great men with enormous influence on Heytesbury—dead within the same decade. Where would we go from here?

CHAPTER SIXTEEN

Second Empire

Heytesbury Stud's lengthy transition between empires ended when Paul took control of the stud in 2000. He emerged from the shadow of his father with a confidence and knowledge of the racing bloodstock industry that mirrored Robert's in quite an uncanny way.

For years Heytesbury had fulfilled Robert's dream, widely acknowledged as it was as 'the best thoroughbred stud in the southern hemisphere'. But for the past ten years, it had been without a leader of Robert's calibre. Was this about to change? Could Paul provide the leadership, the industry knowledge, the family interest so vital to a venture of this nature? Did he have the enthusiasm necessary to propel the stud back to the top of its field?

By now our three legendary stallions—Silver Knight, Haulpak and Pago Pago—had all died. Our foundation sire Silver Knight had lived for three years longer than his stud mates before his death in 1992 at twenty-five years of age.

For the 2000 season, we had three stallions ready to stand at Heytesbury: our resident stallion Carry A Smile (a Haulpak colt from my much loved Silver Smile), the chestnut Royal Abjar who was later sold to Turkey, and Citidancer for the season—but a new bloodline was what the stud needed.

Was it irony or was it foresight that brought the shuttle stallion Second Empire to Heytesbury? What caused Paul and Bob Peters to initiate negotiations with Coolmore Stud in Ireland to bring out the Irish champion? Second Empire had it all: pedigree (by that international sire of sires Fairy King), outstanding conformation, and a brilliant track record. As a yearling, he had cost what appeared to be a remarkable 640,000 guineas, but over the next couple of years, two indicators certainly justified the price. Firstly, there were his sterling performances on the track: as a two-year-old he was unbeaten in Ireland and second top-rated in Europe; he followed this up as champion three-year-old miler in Ireland the following year. And then there was the price that a yearling-colt sired by his half-brother fetched in the Irish sales: a not-often-bid $6 million.

To have a stallion of this calibre standing in Western Australia was certainly a coup for the state, and it didn't harm Heytesbury's reputation either. Second Empire's arrival made Heytesbury the first, and only, stud in the West to have a shuttle stallion. Attended with devotion by his handler Christoph, who was alert to every detail where his stallion was concerned, he settled happily and quickly and his book soon filled to capacity.

Heytesbury began to function as a stud once again. Paul's influence was evident from the first.

Then there was Johann's Dream, the filly that, thanks to Ronnie's thoughtfulness, Paul and I shared. The time had come to try her out on the track. Not knowing quite what to expect, but hoping hard, we raced her in the Heytesbury maroon and white racing silks for her debut at Ascot racecourse in Perth. To our great delight, she proved herself instantly not only by winning first-up, but also by breaking a twelve-year Western Australian class record. A journalist from *The West Australian*, sensing a champion in the making and seeking information on her bloodlines, cut into our jubilation and I was quite astonished by the amount of knowledge Paul had already assimilated. His words could have been Robert's as he responded to the interview.

'She has a blend of Sadler's Wells and Vain bloodlines,' Paul said, 'and this breeding combination suggests she should be better next season. It could also be expected, on pedigree, that she'll win over longer distances.' Even at that early stage he predicted that she would be 'a valuable asset in our breeding plans'. If I had closed my eyes I could indeed have been listening to Robert.

On that occasion I remember Paul corrected me when I said that Johann's Dream was his first racehorse. I was forgetting his first broodmare, Snudge, given to him by Robert, which he had put to Haulpak. The result of that mating was a filly named Vegas Vixen which was put into the care of trainer Len Morton and turned out to be a consistent winner. Out of the 10 times she raced in her first season she was never unplaced. She won at Belmont, Pinjarra and Ascot and her progeny is still prominent on track and in trials.

Why had Paul at twenty-six years of age decided to take control of the stud? I think there were a number of reasons.

Horses and riding had been at the core of his young life. He had started riding early and he must have absorbed his love of horses from his father and I. No doubt he had heard Robert talking about bloodlines and broodmares from the moment he could walk. He'd had his mare Snudge from the age of nine and although his schooling and studies took him away from the stud for many years he'd always had one broodmare or another to call his own.

Increasingly he had begun to feel the difference in the atmosphere at Heytesbury. It was not difficult to see the way the stud was deteriorating both as a business and as a property and he began to realise that since Robert's death what had been lacking more than anything else was family input.

Also he was the only Holmes à Court youngster in the country at the time. Although Peter was in the process of selling his theatre business and his apartment for a move to Sydney later in the year, at this stage he and Divonne were still living in New York with their twin boys, George and Robert. Catherine, Robert and their daughter Emma were living in London. Simon and Katrina had entered a frenetically busy phase of their lives. Katrina's singing career had burgeoned to the point where she was performing professionally. She sang at six operas in nine months and with the associated demands for practice and lessons, the dress rehearsals and performances, in addition to her role of wife and young mother to William, she was physically and mentally exhausted. Simon's schedule was no less intense. His work for i-drive—a computer software company in San Francisco—was not only tremendously exciting, it was valuable, too, in that the innovations with which he was so closely involved placed him at the forefront of new business methods. However, it also involved a punishing work schedule. Every day for months he arrived home at 3 or 4 a.m. only to be back at his desk by 10 o'clock the same morning.

Paul was just as busy with study and business combined. But he was on the spot. He could see firsthand that something needed to be done if the stud were to survive as a stud.

One further influence on Paul's decision was, of course, Ronnie. Although Paul did not take over the running of the stud until after his death, I think Ronnie had suspected that it would happen. His suggested acquisition of Johann's Dream came just at the right time to give my grandson the practical experience he needed in owning a racehorse and a first-hand knowledge of the expenses involved weighed against the possible, sometimes uncertain, gains. After training, stabling, shoeing, vet fees and travel, an occasional win doesn't balance the books. Worse still is the wear on both purse and psyche when your horse is in supreme condition ready to run the race of its life…but pulls up lame. In one interview Janet was reported as saying that racing is 'like standing on a street corner tearing up $100 notes'. Certainly it is not a way to mint $100 bills.

These were among the factors that influenced Paul's decision, but perhaps the reason for his subsequent and almost instant success with the stud came from another direction. As had his father before him, when Paul returned to Australia having graduated from Middlebury, he decided to run a restaurant. He bought the old fire station on Stirling Highway in Claremont, renovated it and turned it into The Station Café. I firmly believe that this equipped him, as it had Robert, to deal with people. Running a restaurant demands an almost super-human level of people skills. Paul had to handle all sorts of situations: drunken cooks, drug addicts, building contractors. He had to learn how to make decisions, resolve conflict, keep the books, make a profit. His father had had similar experiences at La Corvette in South Africa; Paul conducted his business apprenticeship in the same way.

The next thing I knew he had made the decision to sell his business and take over the ailing stud.

If buying the farm nearly thirty years before and turning it into a stud had been a dream come true, Paul stepping in at that juncture was a paddock full of answered prayers. With a diplomacy that outclassed any I have come across he made the changes necessary at the stud.

Slowly the staff situation settled down. Everyone was given a title and an increase in pay. People stopped leaving and started enjoying their jobs again. Christine Boase who had been with Heytesbury through the whole of this transitional phase continued as administration manager; Neale Bruce from Widden in the Hunter Valley, who had worked under the old regime for 18 months, was given the title of stud manager. Tall, good-looking Candice Mitchell was named senior stud hand before being sent to Ireland for the stud season to absorb all she could of new foaling and stud techniques; invaluable Jacque Meyer was night foaling supervisor as well as being my right-hand person. Paul's concerns lay squarely between the well-being of the horses and the job satisfaction gained by the people who looked after them. 'The horses are continuously checked. Staff members are given enough flexibility during the day to make sure all are happy and content. Basically, we don't like to take shortcuts,' he said in one interview with *Racing Western Australia*.

Gradually the stud returned to normal, buying and breeding top broodmares until numbers rose a little again to around forty home mares plus those on permanent agistment.

Meanwhile my beloved Silver Smile had returned to Heytesbury. This was the mare sired by Silver Knight that I delivered in 1975. After her racing days, she had been put to stud and I delivered many of her foals which were always overdue. I remember the concern shown by an American vet who had come out for our stud season one year when she failed to foal on time. I assured him all was well and that she had a custom of always carrying her foal for about a year. During the stud's stormy years this dear mare was sold like so many that were getting on in years. Last stud season, however, her owners were disappointed that she would not conceive so they decided to retire her. Imagine my delight when Paul suggested that the mare return here to her old home.

She was in a shady paddock grazing quite happily behind a grove of trees when I came to the gate. Would she remember me? Rather tentatively, I called her by name and she lifted her head, wondering where she had heard that voice before. How thrilled I was when she trotted to me. We gave each other a great welcome and she was rewarded with the carrots that she had not forgotten I always carried.

Now she is in a larger paddock with several other mares

including her old friend Nolasque and no matter how far away from the gate she is, she will hear my call and come to me…or to the carrots. Nolasque meanwhile has taken on another career at the grand age of thirty. She serves as a nanny to orphan foals, just like the Percheron mares I noticed in Kentucky, and a great job she makes of it too.

Paul has clear and strong ideas on the future of the stud industry in Western Australia. In a recent magazine interview, he said: 'I believe the future is for several studs to share quality stallions. If a number of breeders combine we can get some serious stallions into WA, as New Zealand breeders have done.' Like his father before him, he stressed, 'If WA horses are to compete in this new extra competitive marketplace, we are going to have to produce better and better horses. We can only do this with the introduction of better and better bloodstock into the state.' The answer to that is collaboration between state breeders to enable them to acquire 'access to the best bloodstock in the world and to achieve this we need to pull together'.

I love all my grandchildren deeply. All four are so totally different from each other…individuals with different temperaments, desires and needs which have led to such varied careers. But they do share one attribute—the drive to succeed—and I'm sure this is at least partially a result of being the children of such passionate achievers as Robert and Janet. But whatever the reason, they are all winners in their own right and have proved it by their business acumen; Catherine has added the fierce world of top-class sport to her list of accolades.

The fact that one of the children has decided their career lies with the stud means so much to me. After a long difficult time Heytesbury has the type of leadership it needs. Once again the stud can reach for the stars; can breed its second Cup winner; can yet again become 'the best thoroughbred stud in the southern hemisphere'.

Paul's involvement also means we have a common interest, that we share a language. Again I can communicate my hopes and concerns with someone who understands what I am talking about. Again I can discuss my mares Mercurial Madam and her full sister Wild Rumour. Both have been winners of many races including the Champion Fillies race that Mercurial Madam won in Perth. And when they were retired from racing, both had foals that fetched good prices at the Yearling Sales in Perth and the Easter Sales in Sydney. Mercurial Madam's colt by Anabaa (USA), for example, was sold for $200,000. This is where some of those $100 notes start coming home—selling the colts and fillies from top mating combinations.

Robert built the first empire. The standing of Second Empire at Heytesbury aptly foreshadows the creation of another powerful era for the stud.

CHAPTER SEVENTEEN

'Just Hark at Him…!'

A few years ago Janet approached me with the idea of running coach tours of the stud. She suggested we use the tours both to promote the stud and raise money for one of the charitable organisations with which I am involved. For my part, I was happy to provide a commentary during the forty-minute tour and decided that we wouldn't charge, but that the owners of the coach could arrange a small donation to charity if they wished.

As it turned out, the tours were a huge success, not only in promoting Heytesbury, but in providing an insight into the unique life of a stud for people who know little or nothing about the industry. Visitors get to view the stud and understand the layout of the property, and leave with a great deal more knowledge of what is entailed in the business of breeding horses.

It's usually in the serving barn that the questions really flow. Here's a sample of what they ask.

How many mares a season does each stallion serve?
It varies considerably. Shuttle stallions may serve more than 200 mares a year: well over 100 in the northern hemisphere, and about 100 during the season in the southern hemisphere.

'Go on? Lucky devil. If I have any say about what I come back as in the next life, I want it to be as a shuttle stallion…' say the men. 'Just hark at him!…ha! ha! ha!' is the rejoinder from the wives. Which is just a little unfair of them I think!

When is the stud season?

From the beginning of September to well into December. The horses' traditional birthday is August 1 in the southern hemisphere. The length of a mare's pregnancy is just over eleven months which means that foals can be born over a five-month period from August 1 through to the end of the year.

Does the stallion need to rest between serves?

My goodness, yes! We won't let him serve again in less than four hours.

Have you ever had such a thing as a reluctant stallion?

Usually they are only too keen. Second Empire could make it here from his paddock beside my cottage in less time than it takes to blink! But yes, occasionally they are more reluctant. Our first stallion Silver Knight had a penchant for redheads. He wasn't too keen on the rest and often had to be spoken to. We have a teaser stallion whose job it is to show interest in the mares and whip up excitement all round. The little fellow we have at the moment is a Welsh Mountain pony called David. Last year, once the season was over, we provided him with his own little mate.

How do you know when a mare is ready to conceive?

The teaser will tell—he knows! But of course it is a lot more scientific than that. These days the mare is examined by a vet who should be able to determine quite precisely when she will be ready for mating. It's extremely important not to 'waste' the stallion. The days are gone when the mares were allowed to run freely in the paddock with the stallion. He is much too valuable for a stud to take a chance on his getting kicked or ruined.

Do the mares object to the checking or pregnancy testing processes?

The old girls are marvellous and just put up with it. It's the young ones that make such a fuss.

How do you know which mare to put to which stallion?

The breeding programme is worked out earlier in the year. That was something I so enjoyed doing with Robert. Once that is decided, when the season approaches, the mares are tagged with each individual stallion's colour code. It goes without saying that you need a clear understanding and knowledge of bloodlines, track records and an eye for conformation. But in addition to all this, the greatest successes come when you have a sixth sense as to what is going to make a winning combination. Robert had it when he put Brenta to Silver Knight. Paul has it. It was his idea to have Johann's Dream served over East this year. She was put to Last Tycoon and we'll bring her back to Perth for foaling-down. I have a good feeling about this mating. I think it might produce either our second Melbourne Cup winner or perhaps even the winner of the Golden Slipper…we've been second in that race, but have yet to win.

How much does it cost to put a mare to stud?

A lot of money! Again, it varies considerably from mare to mare. At Heytesbury in the 2001 season, the fees for our stallions ranged from $3,000 to $8,000. Haulpak's $15,000 fee remains the highest ever to be charged in this state. Fees vary in the different parts of the world, too. For example, in Western Australia a stud would be lucky to get $10,000 per serve, whereas in the East they charge between $30,000 and $40,000. In Ireland it would not be unusual to pay $200,000 to put your mare to a top stallion. However, when there is a chance of getting a sum like $6,000,000 for the progeny the stud fee takes on a somewhat different perspective. We don't charge those sort of fees, or get that sort of price, here. There are other costs beside the stallion fee that need to be taken into consideration—travel, agistment, vet fees, farrier—and they all add up. The fee should carry the guarantee of a live foal.

Is a broodmare put to stud each season?

Yes. She usually comes on heat again within 21 days of giving birth, but they will often miss a season. For those who give birth late in the season, you might not want to wait that long and then they are served when they are in 'foal heat', that is within five to eleven days of giving birth. It varies from stud to stud.

What happens to the foal when the mother is being served?

Oh, the foal is close by at all times, in a special little stall in the serving barn. If her baby is out of sight, the mother goes crazy. The foal isn't too happy either.

How long does a broodmare continue to produce young?

It's really impossible to say. One mare we had, called Young Love, had her last foal when she was twenty-two years of age.

What advice would you give someone who thought they wanted to own a stud?

Go for it! I think it's fabulous. I've been here for so many years and never get tired of it. I'm out in all weathers—perennially fascinated by the birth of a foal and the way the mothers react. The only pity is that more humans are not as caring.

What do you do in your leisure time?

Leisure? What's that? I hope I never find out because, quite honestly, I wouldn't know how to go about handling something like 'leisure time'.

One charity with which I am closely involved and which became a beneficiary of these tours is the Association for the Blind, an organisation that can't do enough for its members. My involvement with this institution came about because by the time he died Ronnie was almost completely blind.

During the war, he had lost an eye, and although he appeared to cope with this without any trouble, inevitably as he aged, the sight in his remaining eye grew weaker. A cataract he'd had removed a few years before had helped a great deal and for a while he was delighted he was able to read without glasses. But then, unfortunately, the eye haemorrhaged behind the retina. A man who is deservedly top of his field, leading specialist Ian Constable, warned us that although he could save the eye with laser treatment a scant five to 10 per cent of vision would be all Ronnie would gain.

For someone who had been a great reader all his life, this was a difficult time. The Association for the Blind offered us a mechanical reader, brand new to the market. We had it installed in Ronnie's den and rather tentatively trialled it. It turned out to be a marvel: a machine which enlarged anything from type to photography and which produced a range of variously sized prints on different coloured backgrounds. In the relief that it offered him, it could not have given more pleasure had it been one of the wonders of the world. Now suddenly he could read again. Unfortunately, the machine always seemed to break down at weekends. But then Hans from the Association would kindly rush down to the stud with a spare.

Another aid I found for Ronnie was a talking watch with a button he could press to get the correct time. Sometimes it gave the time in Spanish, which was a cause for more laughter and jollity than anything else as he did have a knowledge of that language. At least he could count in Spanish!

I wish I could tell Ronnie of the latest machine to be invented. This allows a person to place a magazine or book under the reader and it reads the story aloud. How delighted he would be to know about that.

Barbara and Allen Heighton ran tours for the stud for years. It was their suggestion that they add a small levy to their standard charge and that this would be the charitable component. In this case, the first charity to benefit was Riding for the Disabled so they included Fairbridge—the headquarters for the south-west division of the RDA—in the outing as well as lunch and a commentary all the way. Fairbridge is an extremely interesting place to visit in its own right. It is the site of an early boys' hostel, and several graciously aging buildings and an old church lend the peace and ambience of times long ago to the present.

The tours were immensely popular, always full, and we made a lot of money for RDA. A little later, when the first of three reprints of the paperback edition of my book came out, Allen and Barbara sold autographed books on the bus for me, the proceeds of which also went to the charity. In this way, we made a considerable amount of money in quite a short time.

At the annual RDA championships when RDA clubs around the state come to historic Fairbridge for the occasion, one of my jobs as vice patron is to hand out the prizes. The young people take part in games or dressage events and the level of concentration required to perform even the simplest movement can be heart-wrenching. At the same time the pleasure these youngsters get from riding or simply touching the horses is well worth every ounce of effort. The feeling I get when I see the smiles gradually grow on young faces as their horses are brought out is indescribable.

We are gathered to watch the RDA annual championships. A young woman enters the dressage arena on a smart bay pony. I know her well enough to know that she has a complaint called dyspraxia which means that she finds speaking rather difficult and that her memory is particularly poor. Of course in dressage, speech is not an issue…but the ability to memorise is. There are large-size letters of the alphabet placed at selected points around the arena and in executing a round, the rider has to remember which letter to ride towards and in what order. I have found that Riding for the Disabled audiences are always very good, but on this occasion as the contestant enters the ring, the onlookers become quieter than usual. The only sound is the muffled beat of the pony's hooves through the sawdust.

We are all concentrating so hard that it is a bit of a shock when the public address system crackles and the sound of music suddenly bursts out on all sides. Right on cue the rider goes off to execute a perfect round in time to the music. It takes a while for the applause to die down, but when it does we learn that her father is a music teacher and that he had set his daughter's riding test to music. She may have little memory, but she can certainly ride to the music. She passes her test with style and a great deal of verve. I am reminded of a line of a song by singer/songwriter Tania Kerraghan:

> *I close my eyes and I'm on the wind, I can fly when I ride.*

CHAPTER EIGHTEEN

Yellow Daisies and Long-legged Foals

A nd now my story is nearly told. The season is well under way and everywhere I look signs of busyness and life have returned to the stud. Paul continued as he had started, with decisiveness and diplomacy. Eighteen months after he took over the stud, he was made Chief Executive of Heytesbury, the holding company. For the 2001 season, he brought in another stallion: Lord Dane is from an international-class speed family and the son of the great Danehill. Lord Dane is standing beside Carry A Smile and, of course, Second Empire who arrived from Ireland a few weeks ago with Christoph for his second season.

Heytesbury made headlines again a few days ago when the first foal born of Second Empire in the Southern Hemisphere was helped into the world by Jacque. The little bay filly links the new bloodline with the very foundations of this stud. Her dam is Vintage Silver, a half-sister to Carry A Smile, son of Haulpak. Paul is thrilled that the filly has exactly the same markings as Second Empire's progeny in Ireland. She certainly has an exalted pedigree to live up to.

So many years after they left Australia, the young people are all returning: Peter and Divonne live with their twin boys George and Robert in a beautiful house in Sydney; Simon, Katrina and young Will, together with their two beagles, have decided to swap their frenetic Silicon Valley lifestyle for a home in Melbourne; and while Catherine and Rob with their two children, Emma and Fred, are still living in London, they make regular visits back home.

For the first time in ages they will all be here with their growing families at Christmas time. There will be five of my great grandchildren: Peter's twins, Catherine's Emma and young Fred, and Simon's Will. We have a pony and I am giving riding lessons again. We will have a tree and presents and this year with all the youngsters about Christmas will feel like Christmas.

As I write, Zhivago is coiled on my lap. Lara is on her bean bag in front of the fire. Outside it is a beautiful moonlit night. Three foals are due to be born tonight, two from my own horses, Mercurial Madam and Wild Rumour. Mercurial Madam has done well to get over her injury eleven days ago. At great speed, she dashed into a latch on a gatepost while being moved from one paddock to another, with the result that she ripped a huge three-cornered tear in her shoulder. She had numerous stitches, both internally and externally. The accident would have been bad at any time, but only days away from giving birth, she was understandably upset and miserable. In an equivalent state of anxiety, I procured some fresh, unprocessed honey from a local apiary, and applied it to her wound. It was uncanny how quickly it cleared up. Jacque is out under the lights and I know from the signs of anxiety on her face that she worries a little about this birth. From my desk in the study I can see exactly when I need to slip out to give her a hand. Lara picks up on my thoughts; she is at my side, her head on my knee. Just a moment, Lara, nearly finished, nearly finished.

She yawns, only half-believing me.

Before I finish, I must add another word about Africa. There is an old adage that those familiar with Africa know well—*you may leave Africa, but Africa never leaves you*. The saying hints at something deeper than feeling or emotion; it is something in the blood and yet more than that, too. It is found in the rhythm of a heartbeat that is changed forever by a life or even part of a life spent in Africa.

Despite the deep love and commitment I feel for Heytesbury, I still think of Africa. On my long walks up into the ranges, I occasionally turn to gaze out towards the sea and imagine kopjes instead; teaching the little ones to ride reminds me of my riding school in Rhodesia; my visits to the stud at Wallen awaken memories of the wisteria-covered Cape Dutch architecture in the vineyards of Constantia. And sometimes, in the quiet spaces, inevitably the heart of Africa beats within me. Salome, with whom I launched my first book and who owns the Chobe Travel Shop, writes:

> *It is early spring in Botswana and there are signs of growth everywhere. Soon the first animal babies will start arriving. I just love it. I lay awake last night for a long time listening to the elephants, hyena, jackal and hippo contact calling outside and thinking again how blessed we are to have the pleasure of wild animals around us.*

I breathe again the heady scents and sounds of the African bush and my mind flickers back to the many wild orphans we cared for over the years at Chobe. All had names and some stand out among the many. There was Jeremy, the mischievous monkey I finally took up to Erica; little Amanda, the almost hairless monkey brought to us for rearing after her mother had been killed by lions and which, unlike Jeremy, was so loved by our guests; Jasper the young leopard we tended and released back into the wild which had, I'm sure, come to bid me farewell as he sat right in the centre of my path on my final night in Botswana.

And then my thoughts roll forward in time over several decades to my holiday last year with Ala and Lolly Sussens, the couple to whom I had sold my Chobe River Hotel and who now own Tshukudu Game Lodge near Hoedspruit adjacent to the Kruger National Park in the Transvaal.

❖

When Lolly and Ala bought the Chobe River Hotel and Game Camp after Charles' death in 1964, they renamed it Chobe Safari Lodge. Actually that was a name I had always liked, but the rather arrogant solicitor who registered a name for the hotel in the early days simply put down the first name that came into his head, saying that it could easily be changed. He omitted to mention that it would cost fifty pounds to do so—a great deal of money in those days when every penny was essential to build up our dream.

We had met Ala and Lolly from time to time but never knew them well. They were running the very successful 'Sundowner Cruises' on the Zambesi River and had started the famous 'Flame Lily Tours' in conjunction with Rhodesian Airlines.

The death of Charles had affected me deeply and it was a relief that the sale was both quick and friendly. When I left Botswana in a great hurry with my remaining Siamese family, Ching, and two Alsatian dogs, it was Lolly who helped by driving me to Bulawayo, the nearest large town in Rhodesia, a ten-hour journey from Chobe.

I lost touch with the Sussens, but from time to time I had news of them from mutual friends. Evidentally, after about seven years they sold the hotel and moved to South Africa with their two sons. Eventually they started another project: the unique Tshukudu Lodge in the African bush. Lolly had to work hard as the country had been overstocked, the land overgrazed and there were no redeeming features. However, he knew he could improve the land and put in pastures and dams. When Ala first saw the area, she could not imagine ever living there. Thinking back to my first sight of the farm that would become Heytesbury Stud, how well I know that feeling.

If the conservation of wildlife was important in our day, how that importance has escalated now we are aware of the importance of ecological balance: the delicate interaction between the vegetation, animals, insects and bird life. If that critical balance is ever irrevocably destroyed, we may not have a world to pass on to future generations.

And in the twenty-first century, wildlife has another important role in many of the less-developed countries—it is a natural source of tourism income. My stepson, Simon Trevor, currently living in Kenya, recently completed a film on wildlife to raise money to train and guide the young of every nation in saving the wildlife that could bring money into their various countries. This film is produced in every possible language and is having a tremendous effect. One small African boy is heard to say: 'One day I will be a game warden...'

The Sussens are also internationally renowned for their wildlife projects in which the entire family is involved. Just as we did years ago, they introduce as many orphaned animals as possible back into the wild, many of which can be seen in the bush after their successful transition. Ian, the eldest son, and his wife Sylvia run a successful educational and environmental trail for school groups. They also accommodate guests in a self-catering bush camp. Chris, the younger son is involved in the running of the Lodge and is licensed as a professional hunter. Sonja, his wife, runs a small but well-stocked curio shop and from time to time she assists at the Lodge.

A full thirty-five years after Lolly and Ala had purchased our Hotel, I received a surprise phonecall. They were holidaying in Australia and we spent a delightful day at Heytesbury. As they departed, they invited me to visit them at Tshukudu Game Lodge and only four months later Lolly met me on my arrival at the small airport at Hoedspruit, not far from the borders of the Kruger National Park in South Africa.

I spent the most remarkable week with them in February 2001, a special space in time that I will always treasure. Out of the unprepossessing materials they were first faced with, the Sussens have created a particularly impressive game lodge.

Each day of my visit, we were up at six o'clock each morning to partake of coffee and rusks on the verandah of the main building—a very traditional South African way to start the day— before our walk with the wildlife.

As we left on our walk, Tembo, the elephant, was waiting at the gate; two young lions were brought over from their paddock to join us. It was unique. When we reached the dam, the lions had a drink, sprawled in the sand and played like kittens. Tembo gravely accepted the apples and oranges I offered him. A lethargic snake saw little point in moving from his place in the sun. Spiders nested low in trees. The lions showed off, springing into the trees and down again, playing tag with one of the game rangers, and had to be encouraged to join our walk once more.

I fell in love with Savannah, the beautiful cheetah, which was happy to lounge about on the verandah furniture or the comfortable canvas seats at the swimming pool. When we shared a seat and she put her head on my shoulder and purred I could not believe the noise she created—very much louder than Zhivago! I was also amazed at how coarse her fur was, not at all soft and catlike. I continue to keep in touch with this beautiful lovable creature via email.

There is a saying at Tshukudu: that you go as a guest and you leave as a friend. Well, I went as a friend and left as one of the family.

Complementing that is the warm feeling I hold from being such good friends with people with whom I once did business.

Yes, it is true, Africa never quite leaves you. But at the same time, although I have been born and brought up in Africa and that country is indubitably part of me, Australia is the country of my choice. I have achieved a great deal in both countries, but the pride I feel for the stud and its horses is immeasurable. I feel something very much more than that, too—Heytesbury is home.

Morning has broken by the time I get up from my computer. The monochrome of the night is fast-fading as colour seeps into the day and the clouds start to turn a pale pink. Bird song is stealing through the windows, led by the twittering of a family of exquisite blue wrens that have come to live in the tree outside my front door. Soon the crows will take over and their powerful voices will swamp the rest, but of the eighty-three varieties of birds we have spotted on this property, all sooner or later get their chance to pipe in the morning.

It has been a busy night and three new foals are standing at their mothers' sides.

The sun is just coming up over the hills as I step out with Lara for our walk into the ranges. Signs of spring, a different spring from that of Botswana, are everywhere: buds uncurling almost as I look; small insects hovering; birds darting to the ground and back up again, building their nests one twiglet at a time or thrusting their beaks deep into the honeyed centres of the flowering shrubs. I count six ibis balancing gravely on the edge of one of the circular water troughs. The paddocks are green and lush with feed, and everywhere my gaze rests there is a mare standing proudly over her awkward-looking foal. How soon those downy rumpled coats will give way to racehorse gloss. How patient are the mothers with their demanding young.

Lara races ahead of me until she reaches one of our favourite stopping places, not as high as I would have gone on horseback, not as high as I used to climb with Robert, but sufficient to view the stud. There are signs of movement below as the staff get ready for the six o'clock feeds, but otherwise the scene before me is as still as the air itself with its promise of heat. I visualise the stud as it will be later in the morning with the tractors and the 'mule' moving back and forth, mares with foals trotting alongside being shifted from one paddock to another, one stallion or another neighing shrilly on his way to the serving barn.

From here I can see patches of yellow daisies amid the green and the foals starting to pull at the grass, starting to test their long legs.

Thirty years on and I have never tired of it. And at eighty-six years of age—undaunted still—I know I never will. Heytesbury Stud may be a horses' heaven…but it is my heaven, too.

Notes

❖ Some of the people mentioned in this book may need a word of explanation:

– Robert Holmes à Court was my elder son and the founder of Heytesbury Stud. He died of a heart attack at Heytesbury in 1990. He and his wife Janet have four offspring: Peter (1968), Catherine (1969), Simon (1972) and Paul (1973) all of whom figure prominently in my story.

– Simon was my younger son who disappeared in 1977, as far as we know in the Tsitsikamma National Forest in the Cape Province, South Africa.

– Ronnie's full name was Ronald Asheton Critchley. He was born in 1905 in Scotland and grew up on an estate—the focal point of which was a sixteenth century Border keep with a Victorian mansion added on—in the county of Dumfriesshire. He made the military his career and a distinguished career it was. He was much decorated during the war years and by the time he retired from Princess Diana's regiment, the 13th/18th Royal Hussars, it was as Lt. Col. R.A. Critchley, DSO (Burma), MC (Ethiopia), Distinguished Military Medal Emperor Haile Selassie (Ethiopia) and Insignia of Honour (Zambia).

– Jacque Meyer has been at Heytesbury for sixteen years. In addition to being my companion, for many years now she has held the official position of night foaling supervisor and her work with the broodmares is caring and meticulous.

❖ Where in the text I have used superseded names of countries—in the cases of Rhodesia and the Bechuanaland Protectorate, for example—I have done so because that was what they were called at the time of which I was writing. Since Independence they have been called Zimbabwe and Botswana respectively.

❖ I am indebted to Lynne Tinley for rounding out my knowledge of dung beetles. Those interested in a fuller explanation of how that useful insect came to Australia will find a fascinating account in her book *Drawn From the Plains* (London: Collins).

❖ The term 'Waler' was given to all Australian horses abroad and was used to describe a certain type of Australian-bred horse famed for its endurance. Ronnie used to talk about the 'Walers' in his cavalry regiment of six hundred horses in India. He was convinced they came from Esperance in Western Australia and during our trips down to that part of the state, we often speculated about the vast numbers of horses shipped from there for use in the cavalry overseas. However, in the course of my research for this book, I found this not to be so. Instead, it appears the term 'Waler' was given to these horses because they were bred in New South Wales, their original name 'New South Walers' being shortened to 'Walers'. Possibly Ronnie was confusing Esperance WA with Esperance in NSW. Because they had been bred for the purposes of nineteenth-century Australia—for use by stockmen, pioneers, bushrangers and troops—they were renowned for their strength and stamina and worshipped by their owners. By kind permission of the Elyne Mitchell Estate and Curtis Brown (Aust) Pty Ltd, I quote from Elyne's book *Lighthorse, the Story of Australia's Mounted Troops* (South Melbourne: The Macmillan Company of Australia Pty Ltd, 1978, p.7):

> *The name 'Waler'...is not so often heard now as it used to be, before World War II, or before the Indian*

Army ceased to exist. The name was coined in India because many of the horses used by the army there came from New South Wales. According to an article reprinted from The Bulletin of Military Historical Society of the United Kingdom, the Honourable East India Company, at the beginning of the Indian Mutiny, in 1857, found that they had insufficient horses that were large enough to carry British Cavalrymen, and Lieutenant Colonel Connell of the Bengal Light Cavalry was sent to New South Wales to buy horses. Within two weeks he had bought 250 horses between 15.2 and 16 hands. Another 1 000 were shipped shortly afterwards. These horses were exactly what was needed and they became well-known through the army as New South Walers. The British soldiers shortened this to 'Walers'.

I understand that the 'Waler' was registered as a breed in this country in 1986.